Geotechnical, Geological and Earthquake Engineering

Volume 42

More information about this series at http://www.springer.com/series/6011

Alain Pecker • Ezio Faccioli
Aybars Gurpinar • Christophe Martin
Philippe Renault

An Overview of the SIGMA Research Project

A European Approach to Seismic Hazard Analysis

L'ENERGIA CHE TI ASCOLTA.

Springer

Alain Pecker
AP Consultant
Sceaux, France

Aybars Gurpinar
Izmir, Turkey

Philippe Renault
Swissnuclear
Olten, Switzerland

Ezio Faccioli
Studio Geotecnico Italiano
Milan, Italy

Christophe Martin
Geoter-Fugro
Auriol, France

ISSN 1573-6059 ISSN 1872-4671 (electronic)
Geotechnical, Geological and Earthquake Engineering
ISBN 978-3-319-58153-8 ISBN 978-3-319-58154-5 (eBook)
DOI 10.1007/978-3-319-58154-5

Library of Congress Control Number: 2017940276

Printed on acid-free paper

This Springer imprint is published by Springer Nature
The registered company is Springer International Publishing AG
The registered company address is: Gewerbestrasse 11, 6330 Cham, Switzerland

Le doute est un état mental désagréable mais la certitude est ridicule.
Doubt is an unpleasant mental state but certainty is ridiculous.

Voltaire (François-Marie Arouet, 1694–1778)

Contents

Acronyms

ASCR	Active shallow crustal region
BPT	Brownian passage time
CAV	Cumulative absolute velocity
COV	Coefficient of variation
EMS	European Macroseismic Scale
EPRI	Electric Power Research Institute
FFS	Finite fault simulation
IAEA	International Atomic Energy Agency
IDP	Intensity data point
GIS	Geographic information system
GMC	Ground motion characterization
GMPE	Ground motion prediction equation
GR parameters	Gutenberg-Richter parameters
NRC	Nuclear Regulatory Commission (USA)
PDF	Probability density function
PGA	Peak ground acceleration
PGV	Peak ground velocity
PSA	Probabilistic safety assessment
PSHA	Probabilistic seismic hazard assessment
QA	Quality assurance
RP	Return period
RVT	Random vibration theory
SCR	Stable continental region
SHA	Seismic hazard assessment
SSHAC	Senior Seismic Hazard Analysis Committee
SOF	Style of faulting
SRC	Site response characterization
SSC	Seismic source characterization
US-DOE	Department of Energy (USA)

Chapter 1
Introduction

In recent years, attempts have been made to identify and quantify uncertainties in seismic hazard estimations for regions with moderate seismicity. These studies have highlighted the lack of representative data, thereby resulting in predictions of seismic ground motion with large uncertainties. These uncertainties, for which no estimation standards exist, create major difficulties and can lead to different interpretations and divergent opinions among experts. There is a wide consensus among the scientific and technical community for the need to improve knowledge so as to better characterize and, ideally, reduce the uncertainties entering in the calculation of seismic ground motion hazard.

To address this situation, in January 2011, an industrial consortium composed of the French electric power utility (EDF), the French company AREVA, the Italian electricity company ENEL (**E**nte **N**azionale per l'**E**nergia e**L**ettrica), and the French Atomic Energy Commission (CEA) launched an international research and development program. This program, named SIGMA (**S**e**I**smic **G**round **M**otion **A**ssessment, http://www.projet-sigma.com), lasted for 5 years and involved a large number of international institutions.

1.1 Overview of the Project Organisation

The main objective of the research programme was to establish a framework to be used in the future to produce stable and robust hazard estimates. Better characterization and more stable uncertainty estimation could provide input for the updating of regulations. It was expected that total uncertainties will be reduced by significantly lowering epistemic uncertainty, and subsequently, this research programme would significantly contribute to the following efforts:

- Validate, homogenize and stabilize input databases for seismic hazard calculations;

A. Pecker et al., *An Overview of the SIGMA Research Project*, Geotechnical, Geological and Earthquake Engineering 42, DOI 10.1007/978-3-319-58154-5_1

- Produce accepted and validated methods and calculation tools;
- Reduce total uncertainties;
- Improve confidence in seismic hazard assessments; and
- Foster technical and scientific exchanges among French and other European organizations.

The programme was organised around five Work Packages, as follows:

- **WP 1: Improve knowledge of seismic sources**
  The main goal was to produce a French catalogue of earthquakes that covers both the historical and instrumental periods.
- **WP 2: Improve seismic ground motion prediction**
  The goal was to develop methodologies and analysis tools for predicting seismic ground motion that are adapted to the French context and contiguous countries, and which adopt a realistic representation of aleatory and epistemic uncertainties.
- **WP 3: Improve local site conditions representation (Site response)**
  The goal was to develop methods to determine which sites are potentially subject to local site effects, and to develop appropriate tools to be used in seismic hazard calculations.
- **WP 4: Improve seismic hazard models**
  The intention was to better identify and quantify uncertainties with the goal to reduce them, particularly the epistemic uncertainties. It was intended to validate existing methods, and to explore new directions, for testing probabilistic hazard curves against observations.
- **WP 5: Improve on characterization and utilization of seismic ground motion**
  The studies in this work package were aimed to ensure that results of the overall project fulfil the engineers and designers' needs for the design and operation of various types of facilitics. Its goal was to produce methods and tools for the development of the needed engineering parameter(s) for the earthquake ground motion.

Figure 1.1 summarizes the general framework of study in the five main Work Packages.

To help achieve these ambitious objectives, the project management was organized around four entities (Fig. 1.2):

- A Steering Committee (COSS) composed of the industrial financial sponsors, which is charged with identifying strategic orientations and approval of the technical and scientific choices;
- A management committee (COPIL) composed of the Work Package leaders and the Project Manager;
- A international Scientific Committee (CS) to guarantee high quality scientific research and development; and
- An external committee (SHARP) composed of internationally recognized experts to give the COSS a highly credible scientific assessment.

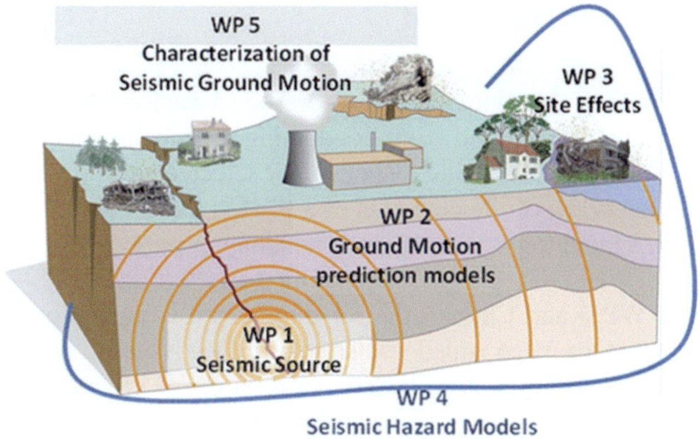

Fig. 1.1 Illustration of relationship between the five technical Work Packages

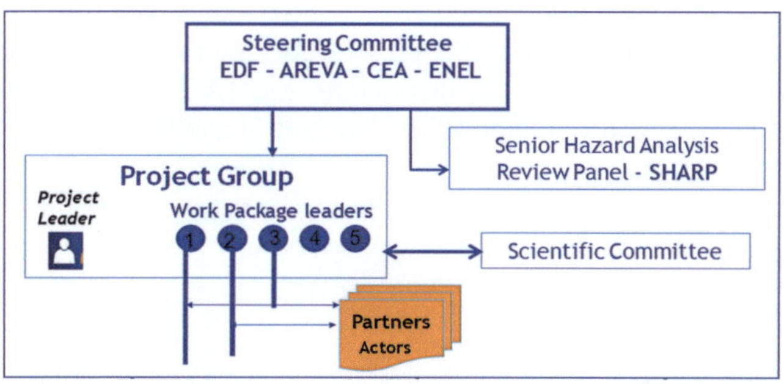

Fig. 1.2 Management flowchart

As mentioned previously, a large number (30) of worldwide academic, research and professional institutions contributed to the project and their contributions resulted in the publication of 75 technical reports reviewed by the Scientific Committee, 40 publications in peer-reviewed journals and numerous communications in international conferences, workshops and symposia.

A list of all institutions and members of the different committees can be found in Annex 1. The most important technical reports (deliverables) mentioned in the present document are listed in Annex 2 with the published papers.

The total cost of the programme amounted to 7.5 million Euro over a period of 5 years.

1.2 Object of the Document

The main objective of this document is to present, based on the outcomes of the SIGMA project, lessons learned from conducting a Probabilistic Seismic Hazard Assessment (PSHA), including site response, for selected areas in France and Italy. After a general overview of the elements of a PSHA, the document is organized in chapters closely related to the work packages: Chap. 3 presents the seismic source characterization (WP1), Chap. 4 the rock motion characterization (WP2), Chap. 5 the site effects (WP3) and Chap. 6 the seismic hazard computations (WP4). Two important chapters have been added related to interface issues to be faced in PSHA between the work packages (Chap. 7) and to the testing of PSHA results (Chap. 8). The final chapter attempts to summarize the lessons learned and to identify the areas where additional research is needed.

It must be stressed that not all the topics related to PSHA were covered in the SIGMA project; nevertheless, they will be mentioned in the document for the sake of completeness.

It is assumed that the reader is familiar with PSHA and, therefore, the basic concepts are not covered in detail in the present document. The interested reader is referred to general documents for further details, e.g.: IAEA Safety Standard SSG-9 (2010), USNRC Regulatory Guide RG 1.208 (2007) and the EERI monograph by McGuire (2004).

References

International Atomic Energy Agency (2010) Seismic hazards in site evaluation for nuclear installations, Specific Safety Guide SSG-9. International Atomic Energy Agency, Vienna

McGuire RK (2004) Seismic hazard and risk analysis, EERI monograph. Earthquake Engineering Research Institute, Oakland

NRC (2007) Regulatory guide 1.208, a performance-based approach to define the site-specific earthquake ground motion. U.S. Nuclear Regulatory Commission, Office of Nuclear Regulatory Research, Washington, DC

Chapter 2
General Concepts and PSHA Background

2.1 Development of a Seismotectonic Framework for PSHA

The first step in building the PSHA model is the collection of geological, geophysical, geotechnical and seismological data from published and unpublished documents, theses, and field investigations. These data are integrated to develop a coherent interpretation of a seismotectonic framework for the study region. Its size can vary depending on the purpose. The international practice for a site-specific study is to distinguish between the investigations at a regional, near regional and site vicinity level (e.g. 300 km, 25 km and 5 km radius in IAEA SSG-9, IAEA (2010)). In order to include all features and areas with significant potential contribution to the hazard, it may also be necessary to include information in a radius up to 500 km (e.g. for subduction zones). This framework provides the guiding philosophy for the identification of seismic sources. Furthermore, the framework should address the important issues that each expert expects to influence the identification and characterisation of seismic sources in the region. The main topics to be addressed in the seismotectonic framework include:

- Use of pre-existing geological structures to provide a basis for defining the present and future seismicity.
- Tectonic models that are applicable to contemporary processes, the observed seismicity, and are compatible with seismic sources.
- Spatial distribution of the seismicity in three dimensions, and associated focal mechanisms and their relation to potential seismic sources.
- Implications of contemporary stresses and strains (e.g. earthquake focal mechanisms, geodetics, other kinematic constraints) for defining sources.
- Use of historical and instrumental seismicity and seismic source delineation to provide a basis for defining the locations of future earthquake activity.

The following categories of seismotectonic configurations can be distinguished:

- Stable continental region (SCR);

A. Pecker et al., *An Overview of the SIGMA Research Project*, Geotechnical, Geological and Earthquake Engineering 42, DOI 10.1007/978-3-319-58154-5_2

- Active shallow crustal region (ASCR); and
- Subduction region.

In Europe two regimes are usually considered and discussed: Active shallow crustal region (ASCR, southern part) and stable continental region (SCR, northern part). Within the SIGMA project the study regions were the South-East of France and Northern Italy. The former region and regime can be considered as being part of the extended crust (SCR) and weighted accordingly, while the seismically active zones of Northern Italy can be predominantly classified as ASCR (see Fig. 4.3). The subduction regime should also be mentioned in this context, but is applicable only for some special regions in Europe, as e.g. Southern Italy, Greece, Turkey and Cyprus. Also, the Vrancea seismic zone in Romania can be treated better with a subduction-related database, although its tectonics are not clear.

The definition of the appropriate seismotectonic model is very important and has implications on many aspects of the PSHA. Especially the source parameters such as the maximum magnitude are closely related with the seismotectonic environment and depend mostly on tectonic metrics (strain rate, etc.). Models for the magnitude scaling and associated uncertainties are also dependent on the defined environment and, thus, have an impact on the hazard.

2.2 Development of Seismic Sources and Logic Trees for Source Definition

Using the seismotectonic framework as a basis, the expert team in charge of seismic source characterization develops its interpretation for the study region (see Sect. 2.5.4). Alternative interpretations of seismic sources (e.g. large regional sources with spatial smoothing of seismicity versus localised source zones) and alternative source zone geometries are usually incorporated in the seismic source models as weighted alternatives using the logic tree methodology. The logic tree framework allows, especially for the seismic source characterization, to capture the epistemic uncertainty lying within the various interpretations. The seismic source zone maps and the supporting calculations of spatial density functions of seismicity, using kernel density estimation, are a part of the seismic source characterization assessment.

2.3 Site Specific vs. Regional Study

PSHA for critical infrastructures (such as dams, power supply structures, e.g. nuclear power plants) is usually done on a site-specific basis and cannot directly be compared to regional studies (such as national seismic hazard maps as used in design codes). The goal of regional studies is to provide seismic hazard results at a regional or national scale based on a uniform approach. Such a result can of course only be achieved if a common seismological rock layer is defined and simplified models are defined in order to keep the computation effort manageable. Usually, the site response cannot be accurately captured in a regional study and, due to the lack of appropriate soil data, cannot be measured in an adequate and accurate way. The seismic source characterization models for regional or site-specific studies can be compared, as the underlying historical and measured seismic data should theoretically be the same. Nevertheless, seismic sources are not always defined through seismicity data. In a site-specific study, the detail of investigation increases as we approach the site, i.e. regional, near regional and site vicinity scales as defined in IAEA SSG-9. Therefore, the sources can also be different from a regional study in which only regional scale tectonic data are considered. On the other hand, the ground motion characterization can also be quite different, since usually no site-specific (or even regional) attenuation model exists. Therefore, the choices for adequate models to be used for the PSHA can be different depending on the targets of the study and the resources allocated to deriving adequate models. For example, in modern PSHA published ground motion prediction models are adjusted to make them more site-specific. Furthermore, recent site-specific studies make use of the single-station sigma concept, which requires some local data and very good knowledge of the investigated site. This is usually not the case for a large scale regional study.

A site-specific study should not primarily rely on the scarce regional data but should undertake the effort to collect adequate near regional, site vicinity and site data at appropriate scales. Such data collection is required by nuclear safety standards (IAEA SSG-9). They are also cost effective and can scale over time depending on the available resources. Without more knowledge and data, the penalty to pay for a site-specific study is the acceptance of large uncertainties. Only site specific data collection to constrain the model space can lead to a reduction of uncertainties.

There is usually a difference in the approach and possibilities for existing versus new sites. At a new site the collection of data for the ground can usually be carried out easily, while at an existing site there are constraints to respect. At a regional level the available data for an existing site might be richer, as equipment has been deployed and measurements have been carried out since then. At a new site in a remote location there might be, in the extreme case, no data at all available, as no infrastructure is nearby. Of course, it depends on the scope of the study and the available resources, but the approach should be chosen according to specific safety objectives and implemented in the context of a long-term perspective. Detailed and extensive data collection can appear costly at the beginning, but will be valuable for reduction of uncertainties and updates at a later stage.

2.4 PSHA – A Framework for Seismic Source & Ground Motion & Site Response Characterization

State-of-the-art seismic hazard studies calculate ground motion exceedance probabilities using earth science hypotheses about the origin and characteristics of earthquakes in the considered study region. Scientific uncertainty about the causes and effects of earthquakes in the study region and about the physical characteristics of potentially active tectonic features lead to uncertainties in the inputs to the seismic hazard calculations.

These uncertainties have to be propagated through the entire analysis. The result is a suite of alternative results (in the form of hazard curves), where each hazard curve is associated with one set of hypotheses and is assigned a weight that represents the relative merit or credibility of that set of hypotheses. These curves quantify the seismic hazard for the study area and its uncertainty at the site, and can be used as basis for decision making. In addition, this suite of hazard curves implicitly contains information about the sensitivity of the hazard results to the various assumptions or parameters and about the contributions of these assumptions and parameters to the total uncertainty in seismic hazard.

Commonly, the method used to calculate seismic hazard at a site is based on the approach proposed by Cornell (1968, 1971). The approach is well established in the literature (Der Kiureghian and Ang 1975; McGuire 1976, 1978) and many studies are based on this mathematical framework. Calculation of the hazard requires specification of the following three main inputs:

1. Source geometry: the three dimensional geographic description of the seismic sources. A seismic source is a portion of the earth's crust associated with a fault with a concentration of historic seismicity or with a volumetric defined region of the earth's crust having similar geological and geophysical characteristics that may be capable of producing earthquakes, thus describing a zone with homogeneous seismic potential (without the ability to identify a precise fault in the source). The geometry of a seismic source relative to the site and a relationship between rupture size and magnitude, m, determine the conditional probability distribution of the distance, r, from the earthquake rupture on the i-th source to the site for a given magnitude: $f_{R(i)|M(i)}(r; m)$. When the hypocentral or epicentral distance, rather than the rupture or Joyner-Boore distance, are used as the distance metric, the distance density distribution is no longer conditioned by magnitude. Most modern ground motion prediction equations (GMPEs) make use of the rupture or Joyner-Boore distance and thus, the distribution of distance is conditioned by magnitude, which becomes especially important for $M > 5.5$.
2. Recurrence: the mean annual cumulative rate of occurrence, ν_i, and the magnitude density distribution, $f_{M(i)}(m)$, of earthquakes occurring in each source i (which can be a specific fault or an area describing a region of diffuse seismicity). This characterisation also includes the maximum magnitude that a source

can produce. Usually, magnitude is characterised in the moment magnitude scale.

3. GMPEs (previously called attenuation relation, function or equation): an algorithm that allows the estimation of ground-motion amplitude (e.g. peak ground acceleration, spectral acceleration or any other seismic parameter) at the site as a function of earthquake magnitude and distance. This characterisation consists of the following three elements: (1) an algorithm for estimating the median amplitude, (2) an algorithm for estimating the standard deviation, σ, that describes the site-to-site and event-to-event scatter in the log[amplitude] of the observations for the same magnitude and distance, and (3) an algorithm for the maximum ground motion that can occur (i.e. an amplitude that has zero probability of being exceeded, given that magnitude and distance). This last aspect has not been considered in SIGMA.

These inputs are illustrated in Fig. 2.1, parts (a) through (c). Figure 2.1a shows the geometry of a seismic source and the distance distribution for a given value of magnitude. The cumulative probability distribution function (PDF) is assigned the form of an exponential distribution, derived on a physical basis from the Gutenberg-Richter relationship; it follows that the density distribution of magnitude, $f_{M(i)}(m)$, for an area source is typically defined by the doubly truncated exponential distribution (Fig. 2.1b). Seismicity for a source with the exponential magnitude distribution is completely specified by the minimum magnitude, m_0, maximum magnitude m_{max}, and recurrence parameters *a* and *b*. Parameter *a* is a measure of seismic activity and parameter *b* is a measure of relative frequency of large versus small events. The log of the cumulative annual rate of events with magnitude m, $\log[\nu_i f_{M(i)}(m)]$, is proportional to bm for $m \leq m_{max}$.

The ground motion is modelled by a ground motion function, as illustrated in Fig. 2.1c. The GMPE is usually expressed in the form $\log[A] = \log[A_{median}(M, R)] + \varepsilon$, where *A* is ground motion amplitude, *M* is magnitude, *R* is distance, and ε is a normally distributed random variable with mean zero and standard deviation $\sigma(M, R)$, that represents variability in log[*A*] for a given magnitude and distance. The maximum ground motion, $\log[A_{max}(M, R)]$, truncates the upper tail of the distribution of ε. For the purpose of the calculations, it is useful to express the attenuation function as the probability $G_{A|M,R}(a^*;m, r) = P[A \geq a^* \,|m, r]$; namely the probability that the ground-motion amplitude *A* is larger than a^*, for a given *m* and *r*.

These three elements (i.e. source geometry, recurrence and GMPE) can be used to calculate the annual probability of exceeding amplitude a^* at the site, which is expressed as the following summation using the total probability theorem:

$$\mathrm{Haz}\left(a^*\right) = \sum_{\mathrm{i}} \nu_{\mathrm{i}} \int_{\mathrm{r}} \int_{\mathrm{m}} G_{\mathrm{A|m,r}}\left(a^*;m;r\right)\, f_{M(\mathrm{i})}\left(m\right)\, f_{R(\mathrm{i})|M(\mathrm{i})}\left(r;m\right) \mathrm{d}m\, \mathrm{d}r \qquad (2.1)$$

in which Haz(a^*), often written as $\lambda(a^*)$, is the mean frequency of exceedance of ground motions, *A*, at a site and thus, the annual rate of earthquakes that produce amplitudes $A \geq a^*$ at the site. The summation is performed over all seismic sources

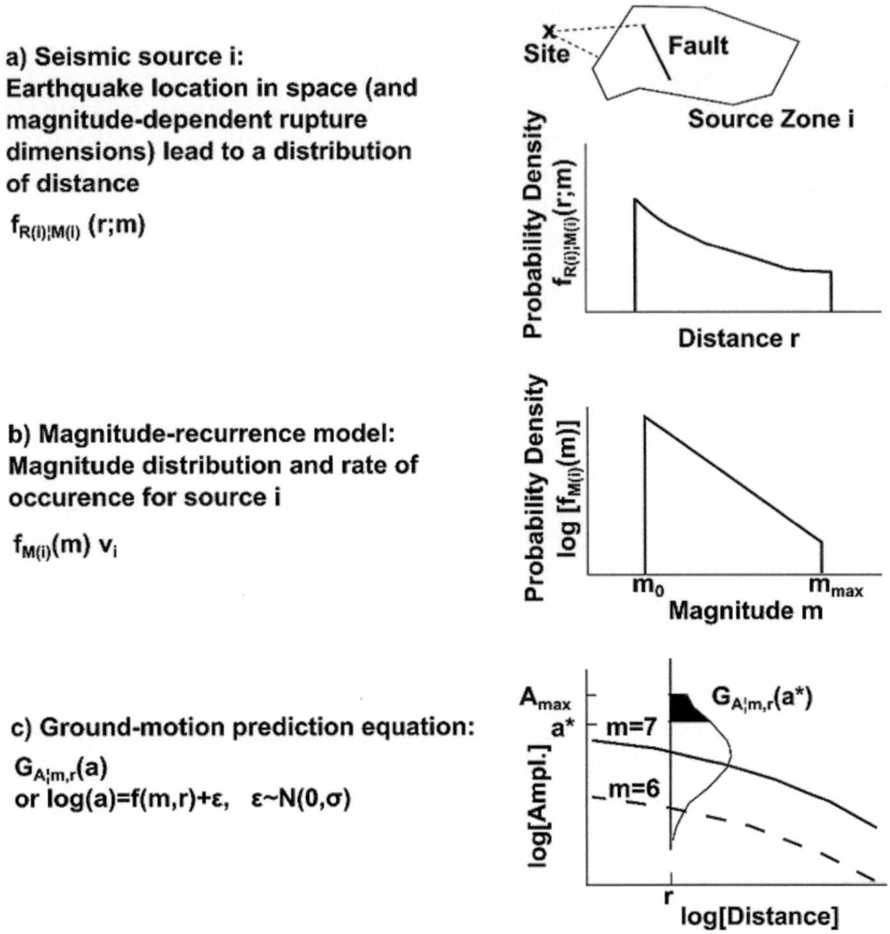

Fig. 2.1 Basic inputs for the calculation of seismic hazard: (**a**) geometry of seismic source and distribution of distance; (**b**) magnitude recurrence model; (**c**) GMPE

i and faults, respectively. The integration on magnitude in the equation considers only earthquakes with magnitudes greater than a minimum magnitude m_0, typically taken as moment magnitude 4 or 5. Smaller earthquakes are assumed to produce no damage to engineered structures, regardless of the ground motion amplitudes they generate (see also Sect. 2.7.3.3).

Both ν_i and $f_{M(\mathrm{i})}(m)$ are typically defined in terms of magnitudes greater than m_0, although lower magnitudes are normally considered in the determination of the rate and magnitude distribution. The hazard equation is formulated using the assumption that earthquakes (most particularly successive earthquakes) are independent in size and location. In most seismic hazard applications, primary interest is focused on computing probabilities for high (rare) ground motions for which the probability of two or more exceedances of a^* in 1 year is negligible. Thus, the quantity on the right side of the equation, which strictly speaking is the annual rate of ground motions with amplitude $A \geq a^*$, is a very good approximation to the probability of exceeding amplitude a^* in 1 year. It is commonly assumed that earthquake occurrences in time are represented by a Poisson random process (Parzen 1962). In fact, this assumption is not necessary, provided the probability of two or more exceedances of a^* in 1 year is negligible.

The calculation in the hazard equation is performed for multiple values of exceedance amplitudes a^*. The result is a hazard curve, which gives the annual probability of exceedance as a function of a^*. This calculation can be performed for multiple measures of ground-motion amplitude (e.g. peak ground acceleration, peak ground velocity, and response spectral acceleration at multiple frequencies). As most GMPEs are formulated for peak ground and spectral accelerations, which are also useful parameters for engineering purposes, these are the usual measures that are used for current PSHAs.

It is useful to understand hazard equation using a deterministic perspective as the starting point. Suppose we want to determine the ground-motion amplitude for an earthquake of known magnitude occurring within a certain seismic source and at a certain distance to the site. It is known that we cannot determine this amplitude exactly, even for fixed magnitude and distance, because the earthquake source cannot be fully described by a single parameter (magnitude) and wave propagation through the earth's crust cannot be fully described by a single parameter (distance). To represent the resulting variability in ground motion, a probability distribution in the form of $G_{A|M,R}\,(a^*; m, r) = P[A \geq a^* \,|m, r]$ is used, i.e. the attenuation equation is written as a complementary cumulative probability distribution. This is simply a method for identifying which of the earthquakes lead to ground motions above the target value a^*.

Suppose now that all potentially damaging earthquakes in a certain seismic source need to be considered. The integral over magnitude and distance in the equation is just a mathematical approach for sampling all possible earthquakes that may occur in the given source, while weighting each earthquake by how frequently it occurs, given the regional seismicity and geology (this weight is expressed by the joint probability $f_{M(i)}(m)\, f_{R(i)|M(i)}(r;m)\; dm\; dr$). Multiplication of this integral by the rate of occurrence ν_i transforms this probability into units of occurrence per year, as required for design decisions and for comparison with other natural and man-made hazards. Note that the most notable distinction between the probabilistic and deterministic character of a seismic hazard assessment is the introduction of the rate of occurrence in the PSHA, rather than just considering and assuming one single scenario. Finally, the summation samples the earthquakes from all of the seismic sources in the region.

Another useful result is obtained if separate "bins" are used to accumulate the rates from earthquakes in different magnitude ranges (e.g. using one bin for magnitudes 5.0–5.4, another bin for 5.5–5.9 and so forth), and then to divide these accumulated weights by the total hazard. The result, which is called the magnitude disaggregation of seismic hazard (McGuire 1995; Bazzurro and Cornell 1999) indicates which magnitude ranges contribute significantly to seismic hazard. Similar disaggregation results can be obtained for distance and for the number of standard deviations (ε). Furthermore, joint disaggregation results can be obtained, where separate bins for different combinations of magnitude, distance, and ε are used (see Sect. 6.5). The consideration and information about ε becomes especially relevant when dealing with lower probability levels, as for example when performing seismic PSA. Furthermore, in terms of hazard computation ε gives an indication about the sensitivity of the result to a potential truncation of the aleatory variability (sigma). This step of disaggregation is also very useful to identify the most contributing seismic scenarios in terms of magnitude-distance couples: this identification is helpful for example to select the appropriate time series for databases of real earthquakes consistent with the UHS (see Sect. 2.7.2.4).

2.5 Logic Tree Approach and Treatment of Uncertainties

2.5.1 Epistemic Uncertainty vs. Aleatory Variability

Modern PSHA studies distinguish between two types of uncertainty, namely epistemic uncertainty and aleatory variability. Aleatory variability (sometimes called randomness) is variability that results from natural physical processes. For example, the size, depth, and time of the next earthquake on a fault and the resulting ground motion can be considered to be aleatory. In current practice, these elements cannot be predicted with sufficient confidence, even with collection of additional data. Thus, the aleatory variability is irreducible without the inclusion of additional predictive parameters. However, an estimate of the sigma, not the sigma itself, is always calculated. On the other hand, epistemic uncertainty (often simply called uncertainty) results from imperfect mathematical models, knowledge about faults and physical processes that produce earthquakes and associate ground motions. In principle, epistemic uncertainty can be reduced with advances in knowledge and the collection of additional data.

Aleatory variability and epistemic uncertainty are treated differently in PSHA studies. Integration is carried out over aleatory variabilities to obtain a single hazard curve, whereas epistemic uncertainties result in a suite of hazard curves based on multiple assumptions, hypotheses, models or parameter values (see Sect. 2.5.2). Results are presented as curves showing statistical summaries (e.g. mean, median and fractiles or percentiles) of the exceedance probability for each ground motion amplitude. The mean and median hazard curves convey the central tendency of the calculated exceedance probabilities. The separation among fractile curves conveys the net effect of epistemic uncertainty in the source characteristics and GMPE on the calculated exceedance probability.

There are epistemic uncertainties associated with each of the three inputs to the seismic hazard evaluation, as follows:

- Uncertainty about the location of causative faults or seismic areas, about the seismogenic potential of faults or seismic areas and other geological features, as a result of (1) uncertainty about the tectonic regime operating in the region and (2) incomplete knowledge of these geological features. There is also uncertainty about the geometry of these geological features (e.g. fault's dip, borders of areal sources, the exact location of a fault, the thickness of the seismogenic layer or alternative interpretations of these geometries).
- Uncertainty in recurrence is generally divided into uncertainty in maximum magnitude, uncertainty in the seismic activity rate ν_i, and uncertainty in parameter b.
- Uncertainty in the GMPEs arises from uncertainty about the dynamic characteristics of the earthquake source and wave propagation in the vicinity of the site. This uncertainty is usually large in regions where few strong motion recordings are available.

Further discussion on the philosophical and practical issues regarding the distinction between epistemic uncertainty and aleatory variability in PSHA is provided by NRC (1997).

Often expert judgment needs to be considered, as the available data are scarce, especially for very rare events with large magnitude. Nevertheless, the importance of data collection needs to be pointed out here, as expert judgment cannot replace real measured data or at least should be guided by it. Conceptually, epistemic uncertainty can be reduced through new data and knowledge which should better constrain the space of alternatives. This should already be motivation enough for each sponsor of a study, e.g. the owners of a critical infrastructure. Furthermore, the collection of local data, which is mandatory for nuclear facilities (see IAEA SSG-9 and USNRC RG 1.208), can help to better understand site-specific phenomena. In the long term this will also enable making use of non-ergodic PSHA models (see e.g. Walling (2009), Walling and Abrahamson (2012)) and, thus, allow for an even more realistic and site-specific hazard assessment.

2.5.2 Logic Tree Methodology

The epistemic uncertainty about the various inputs that affect seismic hazard is organised and displayed by means of logic trees (Kulkarni et al. 1984; NRC 1997). This technique is used for seismic source, ground motion and site response characterisation. For example, each node of a logic tree represents a key seismic source characteristic affecting seismic hazard. This characteristic may be a discrete state of nature (e.g. are identified faults seismically active?) or a numerical parameter (e.g. maximum magnitude on a specific seismic source). In the latter case, the continuous range of values is approximated by a discrete set of values. Each branch emanating from a node represents one alternative interpretation of the source characteristic represented by that node. The collection of all branches emanating from a node is assumed to be a mutually exclusive and collectively exhaustive set of alternative interpretations. The weight assigned to each branch indicates the expert or expert team's assessment of the likelihood that this branch represents the true state of nature, given existing knowledge and data. These weights are conditional on the values of preceding (i.e. lower) branches in the logic tree.

Each end branch of the logic tree represents a complete description of the inputs to the PSHA model presented in the section above, for all seismic sources affecting the site. Associated with each branch tip, there is a weight, calculated as the product of the weights of all branches followed, and a hazard curve, calculated using the hazard equation. These hazard curves, together with the associated probabilities, are used to calculate statistical summaries of the seismic hazard (e.g. mean, median, and fractile hazard curves), and to estimate the sensitivity of the results to the various inputs.

2.5.3 Site Response

The material (soil) beneath a site affects the amplitude, frequency content, and duration of earthquake ground motion at the surface or in embedded layers. From a first-order, engineering perspective, the three most important physical phenomena that affect the amplitude of ground motions at the site are: (1) impedance contrasts between the reference rock used for the rock calculations and the soil medium, (2) resonance effects from energy that is trapped between the surface and the bedrock, and (3) increased damping. In addition, two- and three-dimensional effects are sometimes considered. At high amplitudes of motion, non-linearity may have a significant effect on the elastic properties and damping of the soil.

2.5.3.1 Methods for Implementing Site Response Effects

The most common approach in practice is to perform the PSHA only for rock conditions and then modify the rock amplitudes to introduce the effects of site response. The key disadvantage of this approach is that it does not directly propagate the effects of the aleatory variability and epistemic uncertainty in the amplification factors.

Conceptually, the most straightforward approach for incorporating aleatory variability and epistemic uncertainty in the site response into a PSHA is to start with a site-specific (soil) ground motion equation, which may be obtained empirically or via modelling. Then the hazard equation can be implemented directly for site-specific amplitudes, using these site-specific GMPEs. Alternatively, the rock ground motion model and site amplification model can be treated separately. The advantages of the latter approach are that the required expertise and project workload are decoupled and more combinations of rock motion and site response models are allowed. The disadvantage is that some of the source information available to the rock ground motion model is decoupled from the site response and thus, not available to the site response model (e.g. source location and depth). Various deterministic and probabilistic approaches are used today and SIGMA has investigated and compared some of them (see Chap. 5).

In the past, Bazzurro (1998), Bazzurro et al. (1999) and McGuire et al. (2002) have investigated the accuracy of a number of approximate approaches for the introduction of site response effects in hazard results. NUREG CR-6728 (McGuire et al. 2002) compares several approximate approaches to the direct approach and recommended one (denoted as approach 3) that explicitly includes epistemic and aleatory uncertainty in site amplification, as well as the dependence of site amplification on the rock input motion and on the dominant earthquake magnitude.

This approach integrates over all rock amplitudes, calculating the probability of exceedance of specific soil amplitudes, using means and (log) standard deviations that are functions of magnitude. The resulting equation is:

$$P\left[A_S > a^*\right] = \int\int\int P\left[AF > \frac{a^*}{a}|m,r,a\right] f_{A|M,R}\left(a;m;r\right) f_{M,R}\left(m,r\right) \mathrm{d}m\mathrm{d}r\mathrm{d}a \quad (2.2)$$

where $P[A_S \geq a^*]$ is the probability that soil amplitude A_S exceeds a^*, m is earthquake magnitude, r is distance, a is the rock ground motion amplitude, $f_{M,R}(m,r)$ is the probability that an earthquake equals m and r, $P[AF \geq a^*/a|m,r,a]$ is the probability that soil amplification factor AF exceeds a^*/a given m, r and a, and $f_{A|M,R}(\mathrm{a};m,r)$ is the probability distribution of a given m and r. The formulation recognizes AF as being dependent on m, r, and a and integrates over all m and r to calculate $P[A_S \geq a^*]$. In effect it is doing the PSHA on a rock-modified-to-soil attenuation equation. Bazzurro (1998) found this method to be an accurate way to calculate soil hazard. If, as in McGuire et al. (2002), one recognizes that soil response is governed primarily by the level of rock motion a and magnitude m of the event, the dependence on distance can be neglected, and the previous equation simplified accordingly. A detailed practical oriented discussion of the approaches implemented in SIGMA is documented in Chap. 5.

To evaluate the soil equation (Eq. 2.2), the median site amplification factor (SAF) and the standard deviation of log(SAF) are required. The function $f_A(a)$ is obtained as the negative derivative of the rock hazard curve, and the dominant earthquake magnitude is calculated by disaggregating the seismic hazard.

Epistemic uncertainty in soil amplification is treated by including multiple soil amplification models with weights. For calculation of soil hazard, all possible soil models ($P[A_S \geq a^*|m,a]$ in the equation above) are combined with all possible rock hazard models to calculate a family of soil hazard curves, each curve with its own weight. Statistics on the soil hazard (e.g. mean and fractiles) are determined from this family of soil hazard curves. There are several advantages to using the soil equation over the other alternatives. First, the rock hazard curves can be calculated using region-wide ground-motion equations, rather than developing a set of equations for each site. Second, site-specific amplification models can be derived independently of the seismic hazard study, in the context of soil properties and input motions only. This is how such models are generally applied. Third, this approach allows explicit evaluation of the impact of epistemic uncertainty in soil amplification, which may point to the need for additional data or modelling, rather than combining epistemic uncertainties in soil response with epistemic uncertainties in ground motion attenuation and dependence on earthquake magnitude. Finally, if site-specific amplification models are updated at a later date, for example with additional site data, the soil hazard can be derived (through the equation above) without repeating the entire seismic hazard calculation. Nevertheless, the implementation of this approach should be carried out with great care with regard to consistency among the rock and soil interface parameters. In particular, double counting of uncertainties has been an issue in the past studies (e.g. including the aleatory variability of the ground motion again in the determination of the soil amplification). Recently, the so called single-station sigma concept has been introduced, which attempts to remove the epistemic part of the site response from the evaluation of the aleatory part of the rock ground motion (Rodriguez-Marek et al. 2013).

2.5.4 *Use of Experts*

In order to gather, evaluate and use data in SHA, experts are necessary. Furthermore, to cover the diversity of scientific interpretations, one approach is to involve a team of qualified experts. As a SHA requires a multidisciplinary approach the study makes use of various specialists with different backgrounds and from various fields (such as geology, seismology, geophysics, geotechnical engineering, statistics, risk analysis and computer sciences). The use of experts becomes especially relevant in the context of thc quantification of the uncertainties. In this chapter the criteria for being considered an expert, a very brief example for the expert selection process, and the general process to be followed in eliciting the evaluations of experts are described. Experience has shown that, to be credible and useful, technical analyses such as those performed for the seismic characterisation, ground motion attenuation and site response must: (1) be based on sound technical information and interpretations, (2) follow a structured process that considers all available data, and (3) incorporate uncertainties (see SSHAC, NRC 1997). A key mechanism for quantifying uncertainties is the use of formal expert elicitation. Nevertheless, the term "elicitation" should be used in a broad sense to include all of the processes involved in obtaining the technical evaluations of multiple experts. These processes include reviewing available data, debating technical views with colleagues, evaluating the credibility of alternative views, expressing interpretations and uncertainties in elicitation interviews, and documenting interpretations. In this sense, the evaluation process begins with the first project meeting and ends with the finalisation of the evaluation summaries. The "elicitation" in the context of SHA should not be confused with the classical definition of elicitation used in the social sciences, which is a strictly speaking a poll. Within a study, experts can have sometimes multiple roles (e.g. according to the SSHAC guidelines): they may be acting as proponents and resource experts, as well as evaluators.

2.5.4.1 Guidance Regarding Expert Elicitations

Comprehensive guidance on processes to be followed for expert elicitations has been set forth in the document Recommendations for Probabilistic Seismic Hazard Analysis: Guidance on Uncertainty and Use of Experts by the Senior Seismic Hazard Analysis Committee (SSHAC) (NRC 1997). The guidance was developed under sponsorship of the US Nuclear Regulatory Commission (US-NRC), the Electric Power Research Institute (EPRI), and the US Department of Energy (US-DOE). The study was conducted with the purpose of drawing on the experience gained from expert elicitation projects, particularly those conducted for nuclear power plants in the central and eastern United States, and developing a consensus position regarding acceptable methodologies. In reviewing the differences in PSHA estimates conducted by different groups for individual sites, the SSHAC study concluded that the differences were largely due to procedural differences in the manner in which the PSHA was conducted. Hence, it is concluded that the procedural steps are as important as the technical analyses that comprise a PSHA.

A basic principle defined by SSHAC (NRC 1997, p. 21) is that: "The underlying basis for the inputs [to a PSHA]… must be the composite distribution of views represented in the appropriate scientific community. Expert judgement is used to represent the informed scientific community's state of knowledge."

As noted in NUREG/CR-6372 (NRC 1997, p. 21), the goal of any formal expert elicitation process is: "To represent the centre, the body, and the range of technical defensible interpretations that the larger informed technical community would have if they were to conduct the study." In this context, "informed" means, hypothetically, that all, in the community, have a full understanding of the site-specific technical details. Thus, the experts become informed after having been exposed to the site-specific data and models and having evaluated them. This happens through a number of structured workshops in which data are presented, alternative models and interpretations are debated, and feedback is given. Interaction among the experts should be encouraged and dedicated elicitations should be conducted in form of interviews. Since the publication of the SSHAC guidelines in 1997 a number of PSHA studies have been carried out using this framework and the community has developed some more practical guidance based on the experience gained from those projects (NRC 2012).

Nevertheless, it should be mentioned that there are other approaches to collect expert opinions and integrate them to a PSHA model. The SIGMA project was not carried out under the SSHAC framework but made use of elements of it for building site-specific hazard estimates.

2.6 Interface Issues Between Work Packages

A PSHA is fundamentally a multidisciplinary study that involves various experts from different fields. As the work is often distributed in different work packages that are usually done by different experts or expert groups, the issue of adequate interfaces needs to be addressed (see also Chap. 7).

The source characterization logic trees contain the complete earthquake definitions (magnitude range, style of faulting, depth distribution, dip angles, magnitude-area scaling). The ground motion experts need to consider these definitions when evaluating the candidate ground motion models and address any significant differences between the source and ground motion characterization. Other interface issues to be addressed are, for example, distance conversion relationships, focal depth and mechanisms, which can be related to software implementations.

To improve the consideration of the ground motion and site-response interface, this interface should be supported by a dedicated discussion or workshop. The rock (V_{S30}) to be used for the rock ground motion calculation can be based on the site-specific shear-wave velocity profile at the site. The common understanding of this interface layer, its depth, seismological and soil mechanical properties needs to be developed.

Any interface issues with the end user engineers or PSA analysts should be evaluated at the very beginning of the project in order to ensure that all necessary parameters are addressed and captured. This is especially true with respect to any additional ground motion parameter (e.g. duration, peak ground velocity, average spectral acceleration over a given frequency band) that will be needed, e.g. for the fragility calculations.

SIGMA has tried to consistently address these interface issues in order to improve the hazard estimation. As will be discussed later in the report, SIGMA has, for example, investigated and compared the different approaches to estimate the so-called control point and to assess the impact of the choice of time histories for the site response.

2.7 Common Required Outputs for Seismic Hazard Results

The required outputs depend on the specific intended use of the PSHA. There are different perspectives that deserve to be mentioned and the study and associated output should always be discussed and defined among all stakeholders at the very beginning. First, the purpose of the PSHA could be for design or verification purposes, or for input to a seismic probabilistic safety assessment. Second, the approach will depend on if the output is applied to a new or existing structure, system or component. Thus, the provided requirements below have to be understood as the outcome of best practice and they are only indicative (see also IAEA SSG-9).

2.7.1 *Basic Definitions and Requirements*

2.7.1.1 Control Point

For a site-specific hazard assessment, the ground motion is often presented for the site-specific soil condition at different depths below the surface. Soil hazard curves can, for example, be given at ground surface and foundation level of the turbine building or reactor building, in the case of a nuclear power plant. Commonly, the ground motions at depth are given as outcropping motion. The definition of the project specific elevation levels should be made together with the engineers so as to be consistent with their requirements (e.g. for soil-structure interaction computations).

2.7.1.2 Components of Motion

Usually, the hazard is computed for the geometric mean of the two horizontal components and for the vertical component. This is because most GMPEs use this definition for the horizontal motion. The standard deviation of the horizontal component-to-component variability, which can be used for the development of time histories, can be added back in afterwards if necessary.

2.7.1.3 Vibration Frequencies for Hazard Analysis

As the hazard computation is made for given frequencies, the project needs to define the frequencies required to define the uniform hazard spectrum (UHS). The choice of the relevant frequencies should be made together with the engineers who will subsequently use the results of the PSHA for design purposes or for probabilistic safety assessments. To compute the rock hazard results over a representative frequency range, the following nine spectral frequencies can, e.g., be defined: 0.5 Hz, 1 Hz, 2.5 Hz, 5 Hz, 10 Hz, 20 Hz, 33 Hz, 50 Hz and 100 Hz (which can generally be interpreted as PGA).

The soil hazard should be computed at the nine spectral frequencies given above for the rock hazard plus one or more additional frequencies, so that the site-specific soil resonance is adequately represented. Depending on the number of frequencies that have been used to determine the soil amplification, it may be worthwhile calculating the soil hazard at more frequencies in order to capture the peaks and valleys of the soil response spectrum.

2.7.1.4 Lowest Annual Probability of Exceedance

Today, in order to satisfy the requirements and needs of modern probabilistic safety assessments, the hazard curves should be defined down to an annual probability of exceedance (APE) of 10^{-7}/year (or 10^{-8}/year). The annual rate of exceedance of 10^{-7}/year stated here is a common reference value, e.g. if seismic PSA is necessary for the plant. The actual underlying hazard computation provide much lower probabilities, but the display of results can stop at this lowest annual probability of exceedance for most end users. Nevertheless, it should be noted that the smallest annual frequency of exceedance of interest will depend on the eventual use of the PSHA and the 10^{-7}/year have to be understood as indicative in the following.

2.7.2 Common Hazard Results

2.7.2.1 Seismic Hazard Curves for Reference Rock Site Condition

The reference rock site seismic hazard for the horizontal components for each frequency at a site should be supplied for ground motion levels, e.g. between 0.01 and 10 g and also higher accelerations until the mean annual hazard level falls below 10^{-7}/year in order to achieve a reasonable sampling (e.g. by at least 30 values) of the hazard curve down to low annual probabilities of exceedance.

The range (epistemic uncertainty) in the reference rock site hazard curves should be presented for the horizontal and vertical components for each frequency in plots of the 5%, 16%, 50%, 84% and 95% fractiles and the mean hazard. Numerically, tables of the fractiles at 99 equally-weighted levels should be provided in order to serve as adequate input for the probabilistic safety assessment. The fractile curves are depicted (extended to small annual probabilities of exceedance respectively) until they reach the PGA level where the mean annual hazard has the annual probability of exceedance of 10^{-7}/year. As mentioned above, depending on the final use, the very low annual probabilities of exceedance and fractiles may not be needed (e.g. when defining design values).

2.7.2.2 Seismic Hazard Curves for the Soil Site Condition

The soil site hazard for the horizontal and vertical components for each frequency at each plant should be supplied at the same ground-motion levels as specified in the section above until the mean annual hazard level falls below 10^{-7}/year.

2.7.2.3 Epistemic Uncertainty in the Soil Hazard Curves

The range (epistemic uncertainty) in the soil site hazard curves should be presented for the horizontal and vertical components for each frequency and each plant as follows: Plots of the 5%, 16%, 50%, 84% and 95% fractiles and the mean hazard (supported by tables of the fractiles).

2.7.2.4 Uniform Hazard Spectra

Uniform hazard spectra (UHS) for the horizontal and vertical components for the soil site condition are usually computed and plotted for the following annual exceedance frequencies, provided that no extrapolation of the hazard curves to these levels is necessary: 10^{-2}/year, 10^{-3}/year, 10^{-4}/year, 10^{-5}/year, 10^{-6}/year and 10^{-7}/year. The value of 2.1×10^{-3}/year (= return period of 475 years) can be added if there is an interest to compare the hazard value with that given in the national building code. The ground motions for the UHS are determined by linear interpolation in log-log space between the defined ground motion levels. The range (epistemic uncertainty) in the UHS is shown in terms of the mean and the 5%, 16%, 50%, 84%, and 95% fractiles.

2.7.2.5 Disaggregation

To be consistent with the representation of the UHS, the mean horizontal component rock hazard should be disaggregated in terms of magnitude, distance, and ε (number of standard deviations) at the following levels of annual exceedance frequency: 10^{-2}/year, 10^{-3}/year, 10^{-4}/year, 10^{-5}/year, 10^{-6}/year and 10^{-7}/year. The disaggregation plots (Sect. 6.5) are generated for the following representative frequencies 1 Hz, 5 Hz, 10 Hz and 100 Hz. A disaggregation for additional frequencies can be made on a structure-specific basis. The disaggregation is used to determine the controlling earthquakes in terms of magnitude, distance and ε, which is usually used as guidance to select or develop time histories for engineering purposes. In order to consistently select or develop time histories that are representative of the hazard, the notion of ε is important, as the median prediction is not necessarily the one that contributes the most. Based on ε the engineers will know if an average earthquake (thus, a mean prediction with $\varepsilon = 0$) should be assumed for the site or if the "very unusual" earthquake (thus, above average with $\varepsilon > 0$, e.g. 2 or 3) dominates.

2.7.3 Additional Parameters

2.7.3.1 Ground Motion Duration

Ground-motion durations need to be defined if different earthquake scenarios have to be developed for engineering purposes. There are many available duration models (including global models for large magnitudes and region-specific models for small magnitudes). The models applicable in the context of the study need to be determined with respect to applicability to large magnitudes in the study region (thus, usually with magnitudes up to 8) and consistency with the used ground motion prediction models. Today, the GMPEs used to estimate spectral acceleration (the most commonly used intensity parameter) do not provide a duration as explicit parameter, but they are build including implicitly a duration model based on the earthquakes used to develop the GMPE. The end user of a GMPE expressing spectral acceleration usually overlooks this additional parameter and it has been seen in the past that the regional duration models have a strong effect on the resulting response spectra. Thus, for the sake of consistency between the proposed duration model and the study results, specific thoughts on this issue need to be devoted from the very beginning of a study.

2.7.3.2 Peak Ground Velocity

In the past, there were few ground motion models for peak ground velocity (PGV), so simplified scaling relations between PGV and spectral acceleration were often used to estimate PGV. Many, but not all, of the new ground motion models include models for PGV. The PGV hazard can be computed on the request of the engineers if necessary, but it requires a full re-computation of the hazard for this intensity measure.

2.7.3.3 Cumulative Absolute Velocity

A typical value adopted for the minimum magnitude of the hazard integral is M_w = 5.0 (assumption commonly used for nuclear plants). Based on the observation that the Cumulative Absolute Velocity (CAV) is a parameter much better correlated with the observed damages than other ground motion parameters, the CAV-filtering approach was introduced by EPRI (2006) to consider, in the ground motion assessment, only the contribution of seismic sources having a significant influence on the structures. When such an approach is applied, a lower minimum magnitude is considered (e.g. M_w 4) and the CAV filtering is applied for sources with M_w between 4 and 5.5.

References

Bazzurro P (1998) Probabilistic seismic demand analysis. PhD thesis, supervised by C.A. Cornell, Stanford University, Palo Alto, CA

Bazzurro P, Cornell CA (1999) Disaggregation of seismic hazard. Bull Seismol Soc Am 89(2):501–520

Bazzurro P, Cornell CA, Pelli F (1999) Site- and soil-specific PSHA for nonlinear soil sites. In: Proceedings of 2nd international symposium on earthquake resistant engineering structures. Proc. ERES99, 15–17 June, Catania, Italy

Cornell CA (1968) Engineering seismic risk analysis. Bull Seismol Soc Am 58(5):1583–1606. Erratum: 59(4), 1733

Cornell CA (1971) A probabilistic analysis of damage to structures under seismic loads. In: Howells DA et al (eds) Chapter 27 of dynamic waves in Civil engineering. Wiley, London

Der Kiureghian A, Ang AH-S (1975) A line source model for seismic risk analysis, University of Illinois Technological Report, UILU-ENG-75-2023. University of Illinois, Urbana. 134p

EPRI (Electric Power Research Institute) (2006) Use of Cumulative Absolute Velocity (CAV) in determining effects of small magnitude earthquakes on seismic hazard analyses, EPRI-TR-1014099. Electric Power Research Institute, Palo Alto

International Atomic Energy Agency (2010) Seismic hazards in site evaluation for nuclear installations, Specific Safety Guide SSG-9. International Atomic Energy Agency, Vienna

Kulkarni RB, Youngs RR, Coppersmith KJ (1984) Assessment of confidence intervals for results of seismic hazard analysis. In: Proceedings of 8th world conference on earthquake engineering, vol 1, San Francisco, pp 263–270

McGuire RK (1976) FORTRAN computer program for seismic risk analysis, U.S. Geological Survery, Open File Report 76–67, 69. U.S. Department of the Interior, Geological Survey, Menlo Park

McGuire RK (1978) FRISK: computer program for seismic risk analysis using faults as earthquake sources, U.S. Geological Survery, Open File Report 78-1007. U.S. Department of the Interior, Geological Survey, Menlo Park

McGuire RK (1995) Probabilistic seismic hazard analysis and design earthquakes: closing the loop. Bull Seismol Soc Am 85:1275–1284

McGuire RK, Silva WJ, Costantino CJ (2002) Technical basis for revision of regulatory guidance on design ground motions: hazard- and risk-consistent ground motion spectra guidelines, NUREG/CR-6728. U.S. Nuclear Regulatory Commission, Washington, DC

NRC (1997) Recommendations for probabilistic seismic hazard analysis: guidance on uncertainty and the use of experts, NUREG/CR-6372. NRC, Washington, DC

NRC (2012) Practical implementation guidelines for SSHAC level 3 and 4 hazard studies, NUREG-2117, Rev. 1. NRC, Washington, DC

Parzen E (1962) Stochastic processes. Holden-Day Publishers, San Francisco

Rodriguez-Marek A, Cotton F, Abrahamson N, Akkar S, Al Atik L, Edwards B, Montalva GA, Dawood H (2013) A model for single-station standard deviation using data from various tectonic regions. Bull Seismol Soc Am 103(6):3149–3163

Walling MA (2009) Non-ergodic probabilistic seismic hazard analysis and spatial simulation of variation in ground motion. PhD thesis, University of California, Berkeley

Walling MA, Abrahamson NA (2012) Non-ergodic probabilistic seismic hazard analyses. In: Proceedings of the 15th world conference on earthquake engineering 15 WCEE, Paper 1627, Lisboa

Chapter 3
Seismic Source Characterization

Seismic sources characterization (SSC) relies on the interpretation and integration of a variety of parameters and data primarily collected in a geological, geophysical, geotechnical and seismological database (D4-41, Carbon et al. 2012). It has three fundamental objectives:

- The identification (location and geometry) of all seismic sources contributing to the total hazard at the site of interest. According to the seismotectonic context of the site, and to the considered hazard return period, this involves developing the database within a radius of a few hundred kilometres around the site and to consider different scales of analysis especially when individual faults are considered.
- The characterization of the seismic activity with a large number of uncertain parameters: the maximum magnitude of these seismic sources, their activity rate and the models for seismicity distribution.
- The consideration of all inherent epistemic and aleatory uncertainties and the influence of the different uncertainties in the hazard results.

The main objectives of the SIGMA project were to better understand the sources of uncertainties and their impact on the seismic hazard at the regional scale and at the site scale, and to identify the actions to strengthen the confidence in the mean and median hazard estimates and to reduce the dispersion in the measure of the uncertainty, e.g. in the distance between the lower (5%, 16%) and upper fractiles (84%, 95%) of the hazard curves resulting from the logic-tree.

While the SSC is often seen or considered as the sole responsibility of the geologists, an important lesson gained from the SIGMA project is that the management of the interfaces among the different disciplines is of particular importance to develop a robust and reliable seismic hazard assessment and is a significant indicator of the degree of confidence in the results of the PSHA.

The two PSHA case studies for France and Italy have demonstrated that the development of a complete seismic hazard assessment model requires the intervention of specialists from various disciplines (geology, quaternary geology, seismology,

A. Pecker et al., *An Overview of the SIGMA Research Project*, Geotechnical, Geological and Earthquake Engineering 42, DOI 10.1007/978-3-319-58154-5_3

geophysics, numerical simulation, statistics, soil dynamics, and mathematics). From the identification and characterization of the seismic sources to the modelling of site response, the communication and exchanges between the different specialists and actors involved at each step of the PSHA are not always easy, and the integration of all the components and outputs of the different steps of a PSHA into a coherent model, dealing in an appropriate manner with the uncertainties (e.g. identifying the uncertainties without double or multiple counting and avoiding inclusion of unnecessary branches in a logic-tree), remains a real challenge.

The SIGMA project has clearly identified this issue and was organized in different work-packages with identification of the interactions between them. A significant number of R&D actions were conducted within WP-1 to provide the inputs for the SSC. Some of them were launched with the objective to be directly transferred into operational application (site-specific hazard assessment) while others were more dedicated to academic work with the objective to improve scientific knowledge, to better characterize parameters required in a seismic hazard analysis or to measure their impact on the ground-motion assessment.

The lessons learned during the SIGMA project are considered in this chapter, under two perspectives:

- What are the actions, tasks, efforts that should be considered when conducting a PSHA after SIGMA compared with a PSHA developed before SIGMA; and
- How to identify and manage the uncertainties that influence the hazard results given the objectives of the PSHA.

3.1 Pre-requisites to Develop the SSC Models

At the onset of the SIGMA project, all the R&D tasks of WP-1 and WP-5 were identified by a panel of institutions, research centres and consulting engineers, and implemented by several research teams without possible consideration of the simplifications needed for a PSHA calculation model. These simplifications of the various interpretations and geological models developed by the geologists, provided in the hazard input documents used by the hazard calculation team, are needed to render the PSHA calculations operational.

While it is obvious that not all the research work can be easily transferred into practical application, a site-specific PSHA must rely, especially when developed for highly critical facilities, on unambiguous parameters that describe and reflect the models elaborated by the team in charge of the SSC development. To avoid any misunderstanding between this SSC team and the other teams (Ground motion and hazard calculation teams), the interaction process between them should focus systematically on the preparation of the relevant sections of the hazard input documents with the aim of having a common understanding of the seismic source models to be used for the hazard computation, making sure that there is a common understanding of the seismic source models and of obtaining confidence that the calculation team adequately describes the models and ingredients in a complete and unambiguous way.

To facilitate the interaction between the different components of the PSHA and the consideration of the interfaces, as well as to determine which level of effort has to be devoted to the development of the model, several requirements can be identified in carrying out the future PSHA, in particular:

- Clearly define the objective of the PSHA and the deliverables that are expected to be used by others. The end users are not the developers of the PSHA but most often structural engineers in charge of the design of a new building or the retrofit of existing buildings. If the needs of the end-users are explained, this may orientate the work and tasks to be considered in the development of the models. For instance, if it is known that the structural periods of interest are large, care will be taken by the team in charge of the SSC to focus on the delineation and characterization of seismic sources (and specifically of identified faults) that control the long period content of the ground motion at specific return periods. The objectives of the PSHA study are also a guide to the level of effort in the identification and propagation of uncertainties and in the scale to consider. This is because the treatment of the uncertainties will not be the same whether the aim is to provide ground motion estimates for low or for high probability of exceedance or because the level of efforts will be different if the objective is to produce regional hazard maps or to conduct a site-specific assessment for design purposes. In the last case, the acquisition of data to identify fault sources at the site-vicinity scale, the consideration of local effects (soil amplification, directivity), the GMPEs adjustments, and single-station sigma become standard tasks that are not considered in a regional study.
- Identify the main relevant interfaces between the different components of the PSHA. The interfaces between the various disciplines should be considered at an early stage and at the various steps of the PSHA model development. This is the case between the SSC (Seismic Source Characterization) and GMC (Ground motion characterization) where the actors of SSC must provide the parameters required in the selected GMPEs and vice versa the GMPEs developers must seek, in the development or adjustment of their models, to account for the seismic sources characteristics identified in the SSC model. SSC and GMC developers must also interact with the hazard calculation team to verify that their models are properly included in the PSHA software.
- Conduct preliminary sensitivity analyses to identify the seismic sources and parameters that control the hazard at the site of interest, given: (1) the objectives of the PSHA and, (2) the location and distance between the site and the identified seismic sources. This allows focusing on the resolution of the most relevant components of the PSHA, on the acquisition of data to characterize the controlling sources and generally to simplify the model. As the calculation tools nowadays allow managing complex logic trees, a key lesson learned from SIGMA is that the quality of a future PSHA will greatly benefit from sensitivity analyses and tests that are conducted at the beginning of the PSHA to identify which elements

of the models exert the most significant control on the hazard results, at the specified return periods (and sometimes spectral periods) of interest.

3.2 Database, Earthquake Catalogue, Magnitude Conversions, Uncertainties on Metadata

Prior to the data integration and to the development of the SSC logic tree, the compilation of all existing data and models that are relevant to characterize the seismic context of the site or region of interest, constitutes the first stage of the PSHA. The objective of the compilation is to provide a record and traceability of all the data used during the project, and a description on how the data are interpreted and integrated to develop the SSC models.

As described in D4-41 and D1-27, different scales must be considered from regional to site scale, consistent with the rules or regulation adopted for the project. The database is usually developed at four scales (regional, near regional, site vicinity, and site) as described in the IAEA (2010) SSG-9 guide. When the purpose is a site-specific assessment for NPs or other critical facilities, new data should be acquired or generated through site investigations such as seismic profiles, local seismic or accelerometric networks and geotechnical investigations. For other purposes, new data acquisition would also represent a significant added value for the project.

The SSC model development is based on the coherent interpretation (integration phase) of the data compiled in this database (e.g. existing bibliographic information, databases already developed at regional scales, international publications, Ph.D. theses and new data collected for the project).

A comparison of the PSHA conducted in the two regions considered within SIGMA (South-East quarter of France and the Po Plain in Italy) is relevant to emphasize the impact on the logic trees of new data: the more data are available or newly generated (in case of the Po plain, benefiting from an extensive fault database, see D1-27 and D1-67, dense seismic networks, see D2-72, and in situ measurements), the more the logic tree can be refined and the more new generation approaches (non-ergodic models, rupture simulations, seismic sources modelled as faults rather than as area sources, host-to-target adjustments and single-station sigma) can be implemented, see D4-94. One direct consequence is that the more efforts are paid to the generation of new data, the more it becomes possible to better capture/constrain the uncertainties in the seismic sources and more the predictive models are accurate.

For nuclear sites the seismic hazard assessment must be regularly updated over time and the operator must keep, for audit and safety purposes, a full traceability of all the data (from regional to site scales) collected and generated during the life time of the installation. Consequently, the database should be progressively enriched with the inputs of more local data, like those retrieved during possible additional field investigations, and with the data acquired from monitoring activities during the plant life.

The modern way to develop the database is to use a GIS database, where all the data are georeferenced and all the objects included in the database described by a series of attributes characterizing the geological, geophysical, geotechnical and seismological information.

Keeping in mind the interfaces between the disciplines already mentioned in Sect. 3.1, the database development should not be the sole responsibility of the SSC team but must also include the information pertaining to the ground motion prediction and all parameters that should be deemed necessary by the hazard calculation team.

A description of such a database is available in D4-41. If the information is primarily used to identify any earthquake-related geohazards that may affect the region and to develop the seismic sources models, it must also describe the uncertainties associated with each parameter, to allow the hazard calculation team to quantify the influence of the uncertainties on the hazard estimates.

The databases developed in the two pilot regions (D4-41, D4-94) distinguish the uncertainties respectively ascribed to area sources and to fault sources, which should be considered in any PSHA. For area sources the uncertainties associated with the following are considered:

- The earthquake parameters;
- The boundaries delimiting volumes of the earth crust showing a homogeneous deformation pattern under the present stress field;
- Activity rate/km^2;
- The nature of deformation/style of faulting, considering when necessary single or multiple mechanism of deformation;
- The thickness of the seismogenic crust;
- The maximum magnitude; and
- The parameters of the frequency-magnitude distribution in each source.

When considering fault sources additional uncertainties are considered (D4-94, Ameri et al. 2015) and associated to:

- The 3D geometry of the faults;
- The rupture scenario (segmented/unsegmented);
- The slip rates;
- The characteristic magnitude; and
- The background activity prevailing in the area source where the fault is located.

Assessing the completeness of the database available to develop the SSC models is basically an indicator of the strategy that can be adopted to develop the logic tree and of the uncertainties that can be introduced in the PSHA model. Again, the comparison between the two case studies indicates that the amount of data (existing or specifically collected) is a significant parameter to determine the degree of refinement that it is possible to introduce in the model. In the French case, it was very clear that the limitation of the fault database did not allow developing a fault model covering the entire South-East region with a sufficient level of confidence. The strategy was much more to introduce epistemic uncertainties in area source models considering different criteria to delineate the seismic sources (structural heritage, major domains of deformation or tentative identification of fault systems). On the contrary, the Italian fault database [DISS, INGV; Burrato et al. (D1-27); Valensise (D1-67)] and the new information acquired after the recent Emilia earthquake sequence of May and June 2012, were used to develop a reliable model including fault sources and paying less attention to the uncertainties associated with the area sources.

Among the data, the preparation of the earthquake catalogue (including different time scales) is identified as a crucial step of the SHA because the data are used at several stages of the analysis (determination of maximum observed and assumed magnitude, distribution of activity rates, deformation pattern and so forth). Such a step relies on the compilation, interpretation and integration of multiple sources, including pre-historical, historical and instrumental seismicity.

The strategy deployed between the initial and the refined PSHA in the two case studies was significantly different, even if the common objective was to develop more reliable estimates of M_w. A significant difference in the two regions is that a number of earthquakes of magnitudes between 4.0 and 6.0 occurred in Italy since the 1960s and hence benefit from direct determinations of M_w, while in France most of the earthquakes in this range of magnitudes are only known from macroseismic observations.

For the Po plain, the catalogue used at the beginning of the project was CPTI11 (Rovida et al. 2011), covering earthquakes up to 2007, developed for Italy as the national parametric earthquake catalogue. Sensitivity analyses were conducted in the second version of the PSHA (improvement of the magnitude conversions) and the decision was taken to only consider M_w magnitudes above a minimum cut-off level of M_w 4.5 to avoid magnitude conversions for small events, deemed less suitable to calculate the Gutenberg-Richter parameters.

In the French case, no homogenized catalogue was available at the beginning of the project, each institute developing its own rationale to convert the historical earthquakes defined in MSK intensities and the instrumental earthquakes into M_w. The initial catalogue was developed based on conversions among magnitude scales and encompassed M_w above 2.0 in order to include small magnitudes from which to calculate the Gutenberg-Richter parameters in low activity areas. For the updated model, a new catalogue is considered taking benefit from the national instrumental Si-Hex catalogue in M_w (D1-110; Denieul et al. 2015) and from the homogenized M_w derived from the exploitation of macroseismic intensities and new empirical macroseismic predictive equations (D1-147).

An important lesson from SIGMA (French case) was that the effort for the development of a homogenized catalogue from scratch combining instrumental and macroseismic data was significantly underestimated, the catalogue being only finalized at the end of the project. Significant progress has been accomplished but it remains as an objective to develop a unique catalogue homogenized in M_w, and prepared (with consistency between instrumental M_w and M_w derived from macroseismic data, completeness periods assessment and declustering) to refine the PSHA in the region of interest. More specifically:

- Source parameters have been significantly improved for the instrumental events with the joint effort of the Si-Hex project: EOST developed a relocation and a M_w assessment based on the coda of recorded events (D1-110, Denieul PhD, 2014; Denieul et al. 2015), ISTERRE and CEA made a specific analysis based on cepstral analysis and genetic inversion of teleseismic records to improve the depth estimate of significant events (D1-50, Letort PhD; Letort et al. 2014, 2015). The French catalogue is composed of approximately 40,000 events for which both location and M_w are more accurate. Declustering of the catalogue according to Marsan and Lengliné (2008, 2010) is valuable and can be used as complementary to more standard methods (Gardner and Knopoff 1974; Reasenberg 1985) commonly adopted.
- A preliminary set of source parameters defined for 15 pre-instrumental events in the period 1900–1972 were investigated (D1-129) based on waveform inversion. While the M_w magnitude can be better constrained, the method does not allow for the determination of a robust depth estimate and significant expert judgment is still necessary to infer a best estimate of M_w and depth. The results on M_w represent valuable information on events typically studied only through macroseismic data.
- Exploratory works were conducted on the determination of isoseismal areas using automatic assessment either based on kriging techniques (D1-31 and D1-128) or on manual determination of isoseismals (D1-148). While promising these two alternatives to the method based on calibrated intensity prediction equations, still present limitations (high sensitivity to the parameters adopted in the kriging approach, bias introduced by the expert interpretation of the intensity data points (IDP), no systematic quantification of magnitudes/depth); consequently, it was decided to adopt a more classic and robust approach to refine the historical earthquake catalogue. A preliminary work was also conducted for revisiting the analysis and contextualisation of historical documents (D1-60). This action should certainly be more systematically conducted in order to control the IDP interpretations of the most significant events and to verify that no bias was introduced in the current French historical database, developed by a single analyst since 20 years.
- The definition of macroseismic attenuation models using macroseismic data points to derive M_w for the historical events (D1-108) has required the development of a robust set of well calibrated events including data from the neighbouring regions. A new earthquake catalogue has been developed and its impact on the seismic hazard is still ongoing.

The sensitivity analyses in the Gutenberg-Richter parameters estimates and in the hazard results, conducted using different conversions between M_w and other magnitude scales, as well as on different strategies in the evaluation of M_w from macroseismic data (D1-108), demonstrate that in regions of low to moderate activity these sources of uncertainty in the determination of the correlated parameters *a* and *b* of the G-R magnitude frequency distribution significantly impact the hazard estimates at low annual frequencies of exceedance. This is not observed in the more active Italian region where the G-R parameters are more stable and mainly controlled by the instrumental M_w values. The consolidation of the M_w catalogue in the French area appears as one of the tasks that may significantly contribute to a reduction of the hazard estimate dispersion of the future site-specific PSHA conducted in this region. More efforts should be made in such contexts to collect additional description on the effects of historical earthquakes, to enrich the macroseismic database and to revise the definition of the intensity data points.

3.3 Seismic Source Models

3.3.1 Diffuse Seismicity Versus Identified Seismogenic Structures

Within the SIGMA project, the following two types of seismic sources were considered:

- Area sources where earthquakes occur on faults that are not identified or not identifiable using our current knowledge and understanding of the seismogenic mechanisms. In this case the future earthquakes are assumed to be distributed throughout the entire area source.
- Fault sources where earthquakes only occur on identified seismogenic structures defined by static parameters (geometry of the fault) and kinematic parameters (mechanism of rupture).

To identify the fault sources, it is necessary to exploit appropriate data to develop models representative of a physical process. Since any tectonic earthquake is generated by a causative fault, the concept of area sources of diffuse seismicity zones reveals our inability to identify the causative faults, due either to the impossibility to understand the seismogenic process or to a lack of studies or efforts in acquiring the relevant data. This implies that when fault sources are considered, the seismic hazard calculation focus on the development of models of a physical phenomenon that is constrained by boundary conditions, while when considering diffuse seismicity areas, we model a conceptual process for which the boundary conditions are much looser.

3.3.2 Seismic Source Characterization Framework

One of the objectives of SIGMA was to identify the sources of uncertainty in the seismic source models, to incorporate these uncertainties in the hazard calculations and to quantify their impact and influence on the PSHA outputs.

In consideration of the purposes of the SIGMA project, the conceptual SSC framework was driven by different considerations:

- While a number of R&D tasks have been conducted within the work packages 1 and 4, the driver was much more to focus on individual components of the seismic source models, rather than obtaining new SSC models under SIGMA's ownership. This choice was made because the main priority was given to the identification of the influence of the uncertainties, rather than to the development of new models. This is because the development of these models was considered as requiring a project itself. The method was based on comparisons between PSHA results obtained at the beginning of the project using existing SSC models (D4-29, D4-41), and PSHA results at the end of the project obtained after improvement of components of the SSC models (D4-94, D4-138) on which the R&D efforts focused. Different metrics were defined to measure how the hazard estimates were improved and how much the uncertainties were reduced. This was performed using comparisons between the mean and median of the ground-motion predictor variables as well as ratios between selected fractiles at selected spectral periods and annual probabilities of exceedance of the hazard curves.
- Between the initial and final PSHA runs, refinements of existing models were considered to improve the SSC logic tree. In the French case, a new area source model was developed with the purpose of better accounting for fault systems that were the object of a specific task within SIGMA (Belledonne fault system and faults in Provence) and a fault source model was introduced covering the Provence sub-region. In the Italian case a new area source was identified to account for deeper events and a new composite faults model was introduced.
- The new conceptual models were introduced to consider alternative future spatial distributions of earthquakes (geometry of seismic sources) and the occurrence processes and hazard computation approaches (Poisson using the doubly truncated exponential model and the characteristic model, elapsed time model, renewal model and non-ergodic model – the last applied only in the Po Plain region).

Epistemic uncertainties in the SSC models were initially identified and quantified adopting a process where the teams in charge of the seismic hazard assessment (WP 4) developed their own assessment, based on existing seismic source zonations at national or regional scales, and using as extensively as possible outputs from WP1 as well as inputs produced in other scientific work. The SSC models and identified uncertainties considered in the initial PSHA (D4-29; Faccioli et al. 2012; and D4-41) represent the decision taken and choices by the analysts when integrating the data available at the beginning of the project, developing their interpretations and having alternative solutions in mind. The models were reviewed by independent experts from the scientific committee and interaction with the reviewers and with other experts occurred through the Scientific Committee meetings and critical review of the deliverables. As such, the models initially developed by the two hazard teams were significantly improved in the course of the project (D4-94; Faccioli et al. 2015; D4-138 and D4-169) benefiting from the feedback of sensitivity analyses and the introduction of new alternatives and new outputs resulting from the research tasks based on the recommendations and advice by the Scientific Committee.

3.3.3 Area Source, Fault Sources, Gridded Seismicity

Most of the SIGMA models consider both area source zones and fault sources. The case of fault sources was treated in more detail in the Italian test zone where more geological and seismological data exist. The PSHA models consider, to a lesser extent, also a zone-less approach where the seismicity is spatially smoothed. In such a case the degree of spatial smoothing is controlled by the size and distance of the kernel function. This approach is an alternative to the classical seismotectonic models that consider a well-defined geometry of the seismic sources, and it is of significant interest when there are large uncertainties in the geological model. One drawback of the approach is that the seismicity rates are a function of the kernel distance, which is a difficult parameter to determine. This may lead to overestimating or underestimating the hazard, especially at large return periods, which is a reason why the weights in the logic tree may be different at short and long return periods (D4-170).

3.3.3.1 Area Sources

In the more active Italian area of interest the project principally focused on the development of a composite seismic source model, for which a single area source model was considered as a starting point, i.e. the so-called ZS9 (Meletti et al. 2008) adopted for the official Italian seismic hazard map. This model was refined (D4-29, Faccioli 2013) in the geometry of some of the area sources and, more notably, by introducing a deep dipping zone, subduction-like, that describes the slab under the Po Plain dipping towards the Tyrrhenian Sea, which is justified by the analysis of the earthquake focal depths.

In the French case study, three area source models were considered to capture the variability of the activity rates in a more stable region. All models integrate static and dynamic parameters defined in the GIS database, but consider different strategies to delineate the seismic sources. One model gives more emphasis to the inherited geological structure (Fig. 3.1, D4-41). A second model gives more emphasis to known fault systems and to the seismic activity, as identified by the distribution of

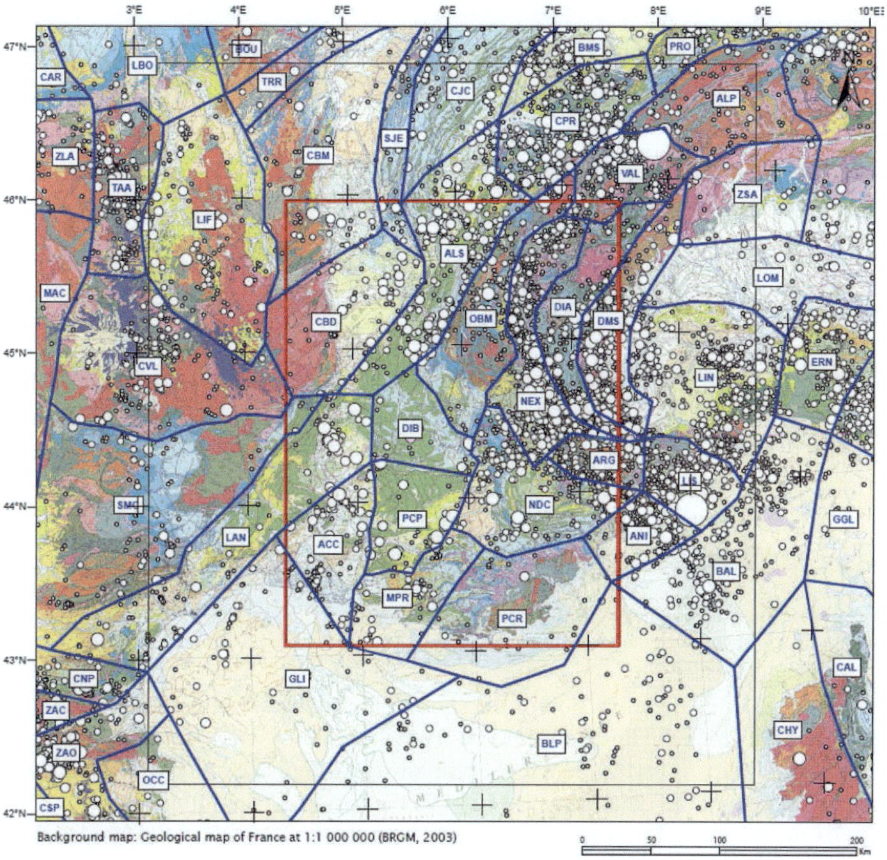

Fig. 3.1 Area source model 1 of the French case study based on the combined interpretation of static and dynamic parameters characterizing the earth crust (D4-41)

historical and instrumental earthquakes. This model considers smaller areas. The third model is more controlled by the identification of a coherent deformation pattern and includes larger area sources.

3.3.3.2 Fault Sources

The consideration of fault source models is of significant interest for sites where fault sources may control the hazard at specific ranges of spectral periods (low or high), when the purpose is to assess the hazard at long return periods, or when directivity effects can significantly affect the ground motion.

For fault sources, the style of faulting and the strike and dip of ruptures may be single or composite. In this case, the logic tree should encompass all types of style of faulting, with weights adding up to 1 (reverse, normal, strike-slip, unknown). The

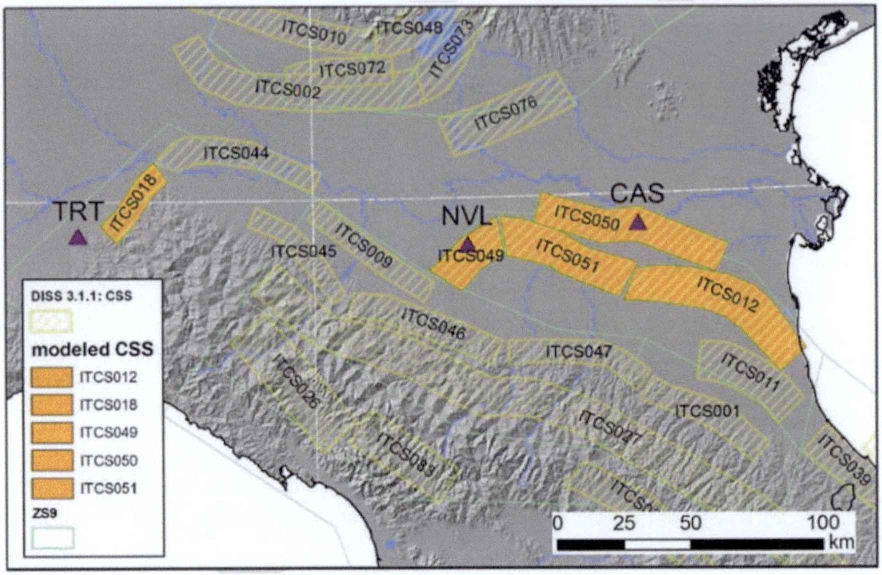

Fig. 3.2 Composite seismogenic fault sources considered in the Po plain (D4-94). *Orange sources* are the fault sources considered for the hazard assessment at three sites (*purple triangles*)

strike and the dip are a function of the style of faulting and of the level of approximation with which the fault is identified from the data.

The reliability of fault source models in low to moderate activity regions is, however, controversial as data to characterize all the parameters required as input to the PSHA model are often missing. Hence, the approaches adopted in the two SIGMA case studies were significantly different.

Composite fault sources were considered in the Italian logic tree (D4-94) with seismic activity described by the characteristic earthquake model for large magnitudes and by the usual Poisson truncated exponential model for the background seismicity. The existing national and regional fault database (DISS Version 3.11, INGV) was used and completed by Burrato et al. (D1-27) and Burrato and Valensise (D1-67) by introducing the data generated by the recent sequence of May 2012, to refine the composite source model (SSC) of the Po Plain (Faccioli et al. D4-94, Fig. 3.2). Because of the limitations of GMPEs to capture site-specific ground-motion properties at short distances from the fault ruptures (like near-field and directivity effects), an innovative approach was introduced by the Italian team. They compare and check, at two sites, the results using a simulation of the ground motion, in which the ground-motion propagation between the ruptures and the site results from deterministically simulated ground motion in replacement of the empirical GMPEs. Finite-fault stochastic simulations were carried out using the EXSIM software (Motazedian and Atkinson 2005) which considers the (1D) local site response, and the generalized description of ground-motion attenuation called GAF (Faccioli 2013, and Faccioli et al. D4-94) was applied to the same composite source model as

5.1.3 Seismic Instrumentation

Seismic instrumentation is understood as installation of instruments that allow recording, on site or in the vicinity, ground motions induced by real earthquakes. Based on these freefield records, empirical approaches can be implemented that could allow to:

- Measure the "site to reference" transfer function or other quantification of this quantity (e.g. coda and duration);
- Consider the use of methods like empirical Green's functions;
- Evaluate the high frequency attenuation of seismic signals (characterized by the "kappa" parameter, see Chap. 4); and
- Understand the potential for seismic activity on nearby faults.

While these approaches are readily applicable in areas of high seismicity (e.g. Japan and Greece), where the data are both numerous and correspond to earthquakes of moderate to high magnitude, they are more difficult to implement in areas of low seismicity, as in mainland France. Indeed, there are few records in these areas and they correspond to distant earthquakes (medium to high magnitude earthquakes far away from the site), or regional and local earthquakes, but with low to moderate magnitudes.

Through continuous measurements made with both velocity meters and accelerometers at several French sites it has been demonstrated that the number of useable events (events with a good signal-to-noise ratio, S/N) obtained during a rather short observation period is high in comparison with what was expected before installing the instruments. At one French site (with moderate seismicity in a rather low noise context) very robust amplification functions were determined with more than one hundred usable events in the intermediate frequency range (Fig. 5.6). On this site, it was also possible to evaluate the κ_0 parameter of Eq. (4.1), which is more difficult to assess because it needs a high-frequency analysis, where the S/N is usually lower.

Even if instrumentation can provide important information about site response, it must be said that it cannot alone solve the whole site effect issue. Indeed, the events that can be expected within a reasonable time interval have usually quite large epicentral distances, which imply that the incidence angle of the wave field is almost vertical. It is, however, worth mentioning that vertical wave incidence is commonly assumed in site response calculations. Furthermore, the recordable events produce (very) weak motions and, hence, they cannot address the non-linearity issues. Nevertheless, instrumental data appear essential to producing empirical amplification functions that are useful when validating numerical simulations for a given range of incidence angles, source sizes and motion amplitudes. Once the simulations are validated with empirical measurements, other scenarios involving extended sources, non-linearity etc. can be explored.

Recommendations have been formulated in Deliverable D3-152 to collect valuable information from the seismic instrumentation; this involves use of velocity

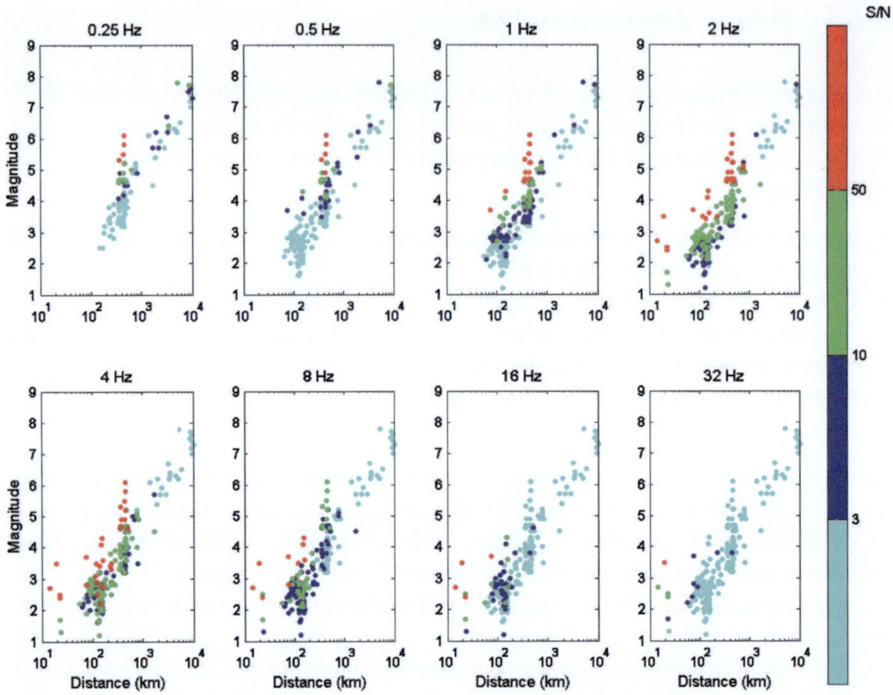

Fig. 5.6 Evaluation of signal to noise ratio for 101 events recorded with velocity meters during 231 days (South-East of France); each subplot presents the analysis at a given frequency

meters, proper choice of the reference (rock) stations locations and the need for continuous recording as opposed to triggered records.

5.1.4 Characterization of Nonlinear Soil Properties

With the presently available technology, nonlinear soil properties can only be obtained from laboratory tests carried out under well-defined and constrained environmental conditions. Needless to say, reliable results can only be obtained if undisturbed samples are used. Retrieving undisturbed samples from the ground in cohesionless soils is a huge challenge that will not be addressed here.

Usually, engineers consider that the nonlinear behaviour is uniquely determined from knowledge of the variation of the secant shear modulus G and equivalent damping ratio ξ with shear strain amplitude: the so-called $G/G_{max} = f(\gamma)$ and $\xi = g(\gamma)$ curves. However, it must be realized that real soil behaviour involves a coupling between shear strain and volumetric strain and that, even under pure 1-D analysis (vertical propagation of shear waves in a horizontally-layered profile), settlements in dry soils or pore pressure build-up in saturated sands may take place. Therefore, a complete description of the soil constitutive model requires data not only on the shear behaviour but also on the volumetric behaviour. Only the linear and equivalent linear models do not require knowledge on the volumetric behaviour because pure shear (vertical propagation of shear waves in a horizontally-layered profile) induces only shear strains.

Even if one focuses only on the shear behaviour, several pitfalls exist that should be properly handled:

- A common mistake in the characterization of the shear behaviour is to measure the G/G_{max} curves in the lab under a given confining pressure and to consider that the same curve applies at any depth in the soil profile provided the material does not change. It is well known, however, that not only G_{max} but also the shape of the curve depends on the confining pressure (Ishihara 1996). To overcome such a difficulty, and to keep the number of tests to a reasonable number, the correct representation is to normalize not only the modulus but also the strain $G/G_{max} = f(\gamma/\gamma_r)$, where γ_r is a reference shear strain (Hardin and Drenvich 1972).
- A difficulty for a complete definition of the shear behaviour, faced during the Prenolin benchmark (Sect. 5.3.3), is the extrapolation of the G/G_{max} curve back to small strains; typically, the laboratory equipment that was used to measure the soil properties was a cyclic triaxial apparatus, which led to inaccurate results for strains smaller than about 10^{-4}. How to reconcile the shear modulus at $\gamma = 10^{-4}$ to the elastic modulus, G_{max}, calculated from in situ geophysical measurements, was a matter of debate and no unique, satisfactory solution was found. This issue is important because it governs the shape of the stress-strain curve in the intermediate strain range, which, in turn, affects the site response. Alternative choices for the extrapolation contribute to an increase of the epistemic uncertainty. This situation can obviously be improved by combining cyclic triaxial tests with resonant column tests, which are able to handle smaller strains.

5.2 Hazard Assessment at the Ground Surface

Seismic hazard at the ground surface may cover different aspects: vibratory ground motion, surface faulting, induced hazards like slope instability, liquefaction and so forth. The SIGMA project only addressed the vibratory ground motion aspect and, therefore, the other topics are not covered in this document.

The most straightforward approach to define the hazard at the ground surface consists of making use, in the PSHA, of generic GMPEs that, through a given proxy, take into account the site characteristics. However, this approach will not give full credit to the peculiar characteristics of the site. One alternative is to start from the rock hazard and to define site amplification functions (SAF), typically described as a function of frequency, which modify the rock hazard spectrum. These site amplification functions are defined by the response spectral ratio of the response at the site divided by the corresponding response at the ideal outcropping bedrock

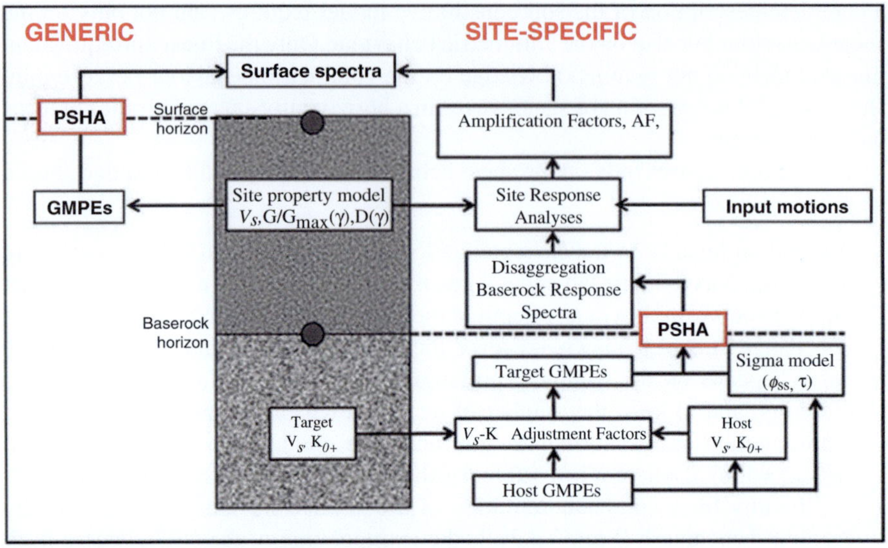

Fig. 5.7 Schematic representation of generic and site-specific hazard calculations for a considered site (Modified from Rodriguez-Marek et al. 2014)

Table 5.1 Classes of approaches to account for site effect in PSHA

Hybrid probabilistic/deterministic		Fully probabilistic	
Generic site (HyG)	Site-specific (HyS)	Generic site (FpG)	Site-specific (FpS)
PSHA at rock + SAF based on seismic norms	PSHA at rock + SAF based on site-specific response analyses or measurements	PSHA based on site-specific GMPEs (single-station sigma)	PSHA at rock + convolution with SAF conditioned to rock ground motion. Typically based on site response analyses

Modified from Faccioli et al. (2015)

(reference station). These SAFs can be either determined experimentally, from observations, or numerically from site response analyses. While in the numerical calculations the availability of a reference station is not generally of concern, since response at an ideal rock site can generally be determined (Sect. 6.7), experimental evaluations of SAFs are in most cases limited by the lack of suitable reference stations in the vicinity of the site.

The two approaches in the ground motion characterization for a site-specific and generic PSHA are illustrated in Fig. 5.7. Note that this figure only considers SAFs evaluated from numerical calculations.

Following Cramer (2003), Bazzurro and Cornell (2004b) and Perez et al. (2009), the classes of approaches to account for seismic site response within a PSHA can be broadly classified as summarized in Table 5.1.

Hybrid approaches are typically based on the results of a PSHA at a rock site, where site response effects are superimposed by multiplying the UHS at rock by a suitable SAF. The latter may be defined either by the spectral amplification factors for generic sites introduced typically by local norms or guidelines (approach HyG), or by a site-specific SAF, calculated in most cases by considering the mean amplification function from 1D linear-equivalent seismic wave propagation analyses for the specific soil profile (approach HyS). In such analyses, time-history calculations are typically carried out by considering a suite of real accelerograms, satisfying the response spectrum compatibility with the target PSHA spectrum on rock. While HyG is the approach implicitly outlined by seismic norms, approach HyS is frequently used for site-specific seismic hazard analyses of important facilities so that it may be considered as the reference approach. Although sound, and easy to understand, from an engineering point of view, a limitation of the hybrid approach is that it may provide estimates of the exceedance rates at the site that are not consistent with the corresponding ones on rock, as noted by Bazzurro and Cornell (2004a, b).

The above mentioned limitations can be overcome by following fully probabilistic approaches, which may be broadly subdivided in terms of their range of application, either for a generic site (FpG) or for a specific site (FpS). The FpG approach is based on the standard application of PSHA, where the site response is summarized within a period-dependent site correction factor to modify the expression of the considered GMPE. Such correction factors are provided by practically all recent GMPEs (www.gmpe.org.uk), either in terms of broad soil categories or in terms of soil classes related to seismic norms, or of other related engineering parameters/proxies such as V_{S30}. The drawback of such an approach is that it may not provide reliable results when dealing with site-specific response evaluations. In the latter case, a site-specific GMPE could be used (Ordaz et al. 1994), if a sufficient number of strong-motion records are available at the site for a reliable GMPE to be constructed, but this is seldom the case. Finally, an FpS approach may be followed, such as described in Sect. 2.5.3.1 and in particular proposed by Bazzurro and Cornell (2004a, b), involving the calibration of conditional SAFs, i.e., of the site-specific ground motion amplification values at a specific vibration period, conditioned to the exceedance of a given level of ground motion on rock.

Although this fully probabilistic site-specific approach allows a formally correct incorporation of seismic site response into the PSHA, it suffers from several limitations that were addressed in Deliverable D3-54, namely:

- the probability distribution of the conditioned amplification function is based on 1D numerical simulations of vertically propagating plane waves at a nonlinear soil site with uncertain properties: this assumption is expected to deeply affect not only the median amplification function, but also its standard deviation;
- observed amplification at the site may also be affected by source-to-site azimuth and directivity, especially in near-source regions (e.g., because of different angles of incidence of waves, or because of larger/smaller onset of surface waves depending on the relative position of the source with respect to the basin). This is neglected by 1D approaches; and

- no validation is available against a sufficiently wide set of strong-motion records, especially to quantify the σ values.

The Bazzurro and Cornell (2004a, b) approach has been followed in recent projects like PEGASOS and PEGASOS Refinement Project (Renault 2009). However, it has not been implemented within the framework of the SIGMA project and, consequently, it will not be presented in detail herein. The interested reader may refer to the original work of Bazzurro and Cornell (2004a, b) or to Deliverable D3-96 in which it is briefly outlined. Within the framework of SIGMA the three other approaches, HyG, HyS and FpG, have been tested by the French and Italian teams. The Italian team applied the method to the Casaglia site in the Po Plain (Deliverables D3-96 and D4-94, a further application to the Mirandola site in Po Plain is illustrated in Faccioli et al. 2015) and the French team applied it to two sites: Grenoble (France) and EuroseisTest (Greece) (Deliverable D4-153).

5.2.1 Direct Evaluation from Ground Motion Prediction Equations (FpG)

Instead of computing the rock hazard and then transferring it to the ground surface with approaches that will be detailed in the following paragraphs, relevant GMPEs can be used in the PSHA to directly compute the hazard at the ground surface. This approach assumes that: (a) the soil conditions at the site resemble those at the stations in the database considered for the development of the GMPEs used for the hazard estimation and (b) the site response is assumed to be correctly captured by the site model included in the adopted GMPEs. To be valid, the GMPE must be representative of the site conditions, i.e. include a proxy that is deemed to represent the ground conditions; some GMPEs may also include a nonlinear term for high-amplitude motions. The most commonly used proxies are the V_{S30} and the site fundamental frequency f_0 or the site class (Eurocode or NEHRP classification); examples of relevant GMPEs are given in Deliverable D3-152. This approach is crude because each site has its own peculiarities like, for instance, interbedded layers with a high stiffness contrast, marked subsurface topography and so forth and use of a single proxy cannot account for all the site-specific features. Nevertheless, it can be argued that empirical GMPEs are established from large databases, which also certainly contain peculiar sites. The main advantage of the approach is that a full probabilistic analysis is possible with propagation of all (certainly overestimated) uncertainties. Note that even in this simplified approach, a good characterization of the soil profile is needed (V_{S30}, soil class, f_0). As mentioned previously, this approach has been implemented for the Grenoble, Po Plain, and EuroseisTest sites. Results will be compared to the other approaches in Sect. 5.3.2.

An alternative site-specific approach in the same broad category (FpG) consists in applying data-based corrections to the GMPEs median values and in replacing their standard deviations. This approach can be implemented when a sufficient number of records is available at the target site. It is also desirable that the recorded earthquakes span a sufficient range of magnitudes, distances and azimuth angles. Following this approach the GMPE's median predictions are modified through site-specific correction factors (δS2S) and the GMPE's sigma is replaced with the single-station sigma ($\sigma_{ss,s}$), as explained in Sect. 4.3. The δS2S factor can, in first approximation, be considered as an intrinsic characteristic of the site and is used to modify the predictions by a GMPE, in a very simple way. More precisely, the site correction term modifies the GMPE median prediction (μ_{GMPE} (T)) as follows:

$$\mu_{\text{corrected}}(T) = \mu_{\text{GMPE}}(T).10^{\delta S2S(T)} \quad (5.1)$$

This approach has been implemented by the Italian team for the Casaglia (CAS) and Novellara (NVL) deep soil sites in the Po Plain. A generic, regional GMPE has been established (in Deliverable D2-72) to a large extent from records on sites in the Po plain with similar characteristics to the Casaglia and Novellara sites; the median predictions from this GMPE were then modified with the δS2S term established from the acceleration records obtained at both sites. Likewise, the original standard deviation of the GMPE has been replaced by the single-station sigma derived from the same records. Results for the Casaglia site are compared to the standard approach (direct use of a GMPE for the relevant soil class) in Fig. 5.8.

The figure below has been established for just a single branch of the PSHA logic tree (one source model, one GMPE) and compares the standard approach (in blue) and the site-specific approach with the above correction (in green). The other curves in Fig. 5.8 correspond to different approaches that are presented in the following paragraphs, whereas the implications of the present approach on the σ terms are discussed in Sect. 5.4. As seen from the figure, the CAS example strongly indicates that the soil surface spectra can decrease significantly when using results from local site analyses (single-station coefficients) instead of the site coefficient of the GMPEs. This behaviour is however site-specific and does not occur for instance at NVL.

5.2.2 *Generic Site Specific Approaches (HyG)*

This approach assumes that the hazard calculated with GMPEs, either at rock or at the surface, is a first-order model for the target site response but that some site-specificity must be introduced in order to provide a better description of the site response. A correction factor can be developed and applied, as a post-processing, to the computed hazard spectrum.

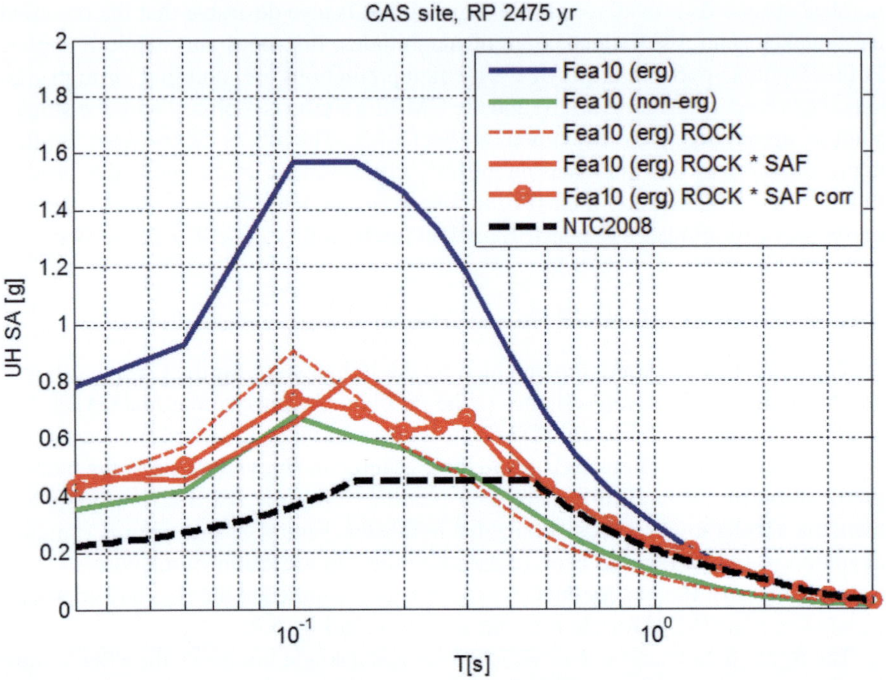

Fig. 5.8 Casaglia (Po plain) site. Site specific median ground surface spectrum for 2475-years return period

5.2.2.1 Correction to Rock Spectrum

The most commonly accepted meaning for this approach consists in starting from the UHS at the bedrock and applying a site amplification factor (SAF), based on some site characteristics, to compute the surface spectrum. SAFs can be given by norms, based on the site class (e.g. Eurocode or NEHRP), or can be measured experimentally on site (ideally in a vertical array) if the latter is located in a reasonably active area.

As neither Italy nor France have a sufficient amount of data from a single site, the approach could not be fully tested. However, taking advantage of the Japanese KiK-net data (http://www.kyoshin.bosai.go.jp/), Paolucci et al. (Deliverable D3-96) calculated the SAFs from stations exhibiting a V_S soil profile similar to those observed in the Po Plain, i.e., deep soil sites with V_{S30} values in the range 200–400 m.s^{-1}. From their study at 21 stations and considering only shallow events (depth < 15 km) with PGA > 10 cm.s^{-2} they reach the following conclusions:

- observed variability of SAFs at KiK-net deep soil sites is generally limited in spite of the wide range of magnitude and distances encompassed by records (Fig. 5.9);

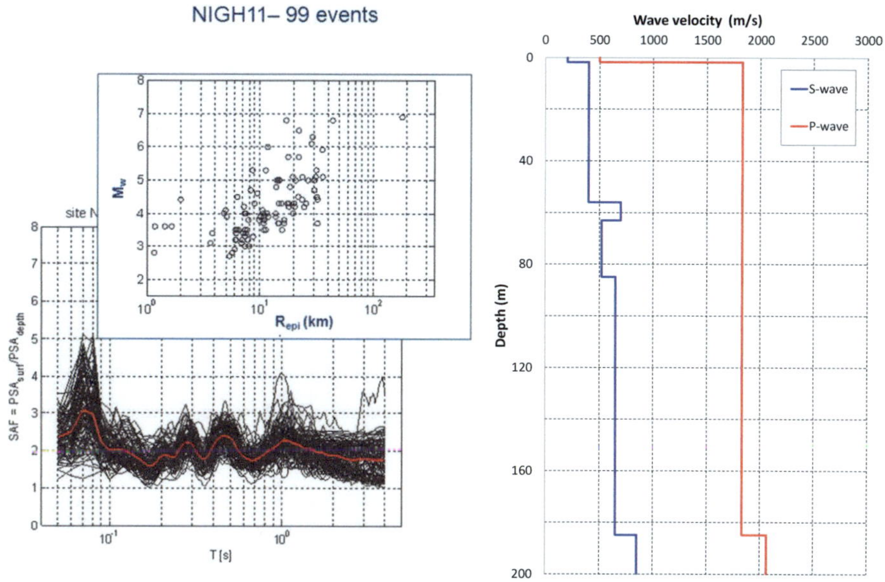

Fig. 5.9 Unconditioned SAFs (*left side*) and Vs profiles (*right side*) for one deep soil site station (NIGH11) in the Kik-net. The M_w-R_{epi} distribution for the considered records is also shown

- conditioned SAFs (i.e. spectral amplifications as a function of the spectral pseudo-velocity on rock, PSV^R) do not show a significant, if any, dependence on the intensity of motion at bedrock, suggesting, for the stations considered, no clear evidence of nonlinear site response, in spite of the relatively soft soil conditions (Fig. 5.10). This is at variance with results from Régnier et al. (2013) who found significant evidence for the shift of the predominant peak frequency with increasing amplitude of motion at the borehole station. Reasons for such discrepancy are not yet understood;
- $\sigma_{\log10}$ computed from the conditioned SAFs ranges typically between 0.04 and 0.08; furthermore, no clear evidence is found of a dependence of $\sigma_{\log10}$ with period.

A similar finding of a limited aleatory variability of SAFs from low intensity records at deep soil sites has been confirmed by the surface-to-borehole spectral ratios observed from a selected set of records at Casaglia. The interesting remark on these records is that the observed SAF remains nearly unchanged, irrespective of earthquake magnitude and distance, both for local weak events and for distant stronger earthquakes. However, in none of the selected cases, was the ground-motion amplitude high enough to predict any significant nonlinear effect.

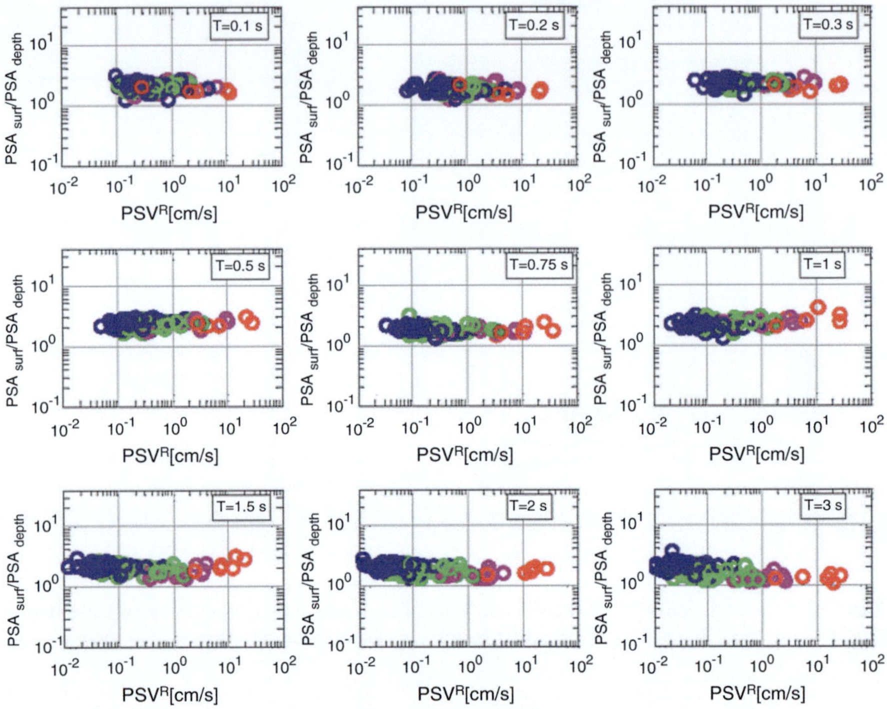

Fig. 5.10 Conditioned SAFs for different vibration periods for station NIGH11, with data grouped by Magnitude. *Blue dots*: $M < 4$, *green*: $4 < M < 5$; *magenta*: $5 < M < 6$; *red*: $M > 6$

5.2.2.2 Correction to Ground Surface Spectrum

The approach suggested here consists of calculating first the ground surface response spectrum with the relevant GMPEs (Sect. 5.2.1), and then of applying a site correction factor on the rationale that the GMPEs do not reflect all site peculiarities, for instance a significant 2D sub-topography (basin effect). The correction factor can be either derived from statistical relationships (Site Amplification Prediction Equation, SAPE) or calculated with reference to a 1D analysis.

SAPE was introduced by Cadet et al. (2012a, b), to define empirical prediction of site amplification as a function of a few parameters (V_{SZ}, with z equal to 5, 10, 20 and 30 m and f_0) derived from Japanese strong-motion data (KiK-net). The amplification factor was estimated from the ratios between the surface and down-hole horizontal response spectra, corrected for the varying depths and impedance of the down-hole locations. The amplification factors were then correlated with site parameters. The results showed that the best performance in predicting site amplifications was obtained by the coupled parameters $V_{S30} - f_0$, while the best single parameter proved to be f_0. The hazard spectrum should then be multiplied by the amplification function. However, care must be taken in the combination of hazard spectrum and amplification function in order not to double count the site effect in both GMPEs and SAPE. For example, a SAPE based on f_0 can be applied to a hazard spectrum calculated for the selected V_{S30} value in the adopted GMPEs, assuming that V_{S30} and f_0 are poorly correlated and thus account for different origins of site amplification. On the contrary if a SAPE based on both V_{S30} and f_0 is used, then the hazard spectrum should be calculated using a reference rock V_{S30} (e.g., $V_{S30} = 800$ m. s^{-1}) in the GMPEs.

The advantage of this approach is to allow considering the amplification at the fundamental frequency of the site, which is usually neglected in the generic approach because only few GMPEs adopt the fundamental frequency as a site-effect proxy. The main limitation in the application of SAPE is related to the database, consisting exclusively of Japanese data, which may emphasize systematic differences in the shallow site amplification in Japan with respect to other regions. Due to this limitation, use of SAPE is not recommended.

To account for sites prone to 2D or 3D amplification (basin effect, above surface topography) much larger than implied in the generic GMPEs, the notion of aggravation factor (AG) was introduced by Chavez–Garcia and Faccioli (2000) to reflect the contribution in site effect due to a complex local geometry with respect to 1D geometry. It is defined as the ratio between 2D/3D and 1D calculated response spectra. AGs may be obtained from 2D/3D calculations with different soil constitutive models (linear, equivalent linear, fully nonlinear) and using accelerograms consistent with the hazard level. Ratios between 2D/3D and 1D response spectra are then computed and the final aggravation factor (and associated standard deviation) is deduced. The AG is then applied to a reference hazard spectrum in which the 1D response of the soil is assumed to be accounted for. AGs are based on ratios of computations: as a first approximation, it is assumed that a change in the model description (e.g. a velocity change) will affect more or less similarly both 1D and 2D/3D computations. The overall results are then less sensitive to changes (inaccuracies) in the model description. AGs have also been proposed, based on an extensive number of calculations and simplified typologies, by the NERA project (http://www.nera-eu.org).

An example of aggravation factor calculated for the EuroseisTest site is depicted in Fig. 5.11 (Deliverable D3-152).

5.3 Completely Site Specific Approaches (HyS)

The site-specific approaches are based on a detailed consideration of the local site response and relative uncertainties. The site amplification is estimated using more or less sophisticated numerical or experimental (Sect. 5.2.2.1 for one example) methods depending on the characteristics of the site. In order to solve this problem several steps need to be considered (see Fig. 5.7):

- definition of an input motion at the base of the soil profile. This may require the calculation of the rock hazard for large values of V_{S30} (hard-rock conditions) that

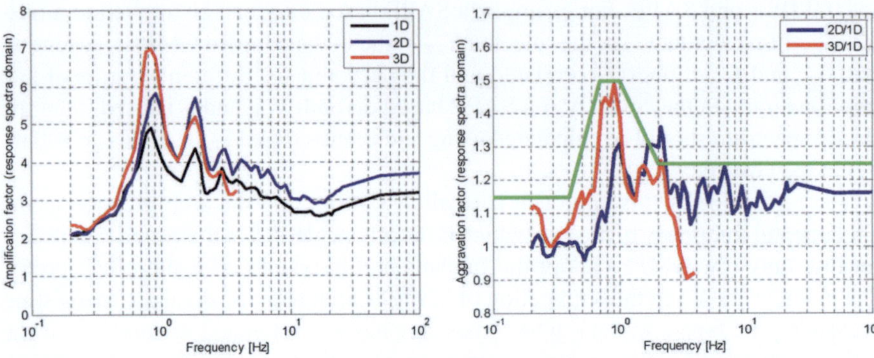

Fig. 5.11 EuroseisTest: *Left* – Calculated amplification factor; *Right* – suggested aggravation factors

may be outside the domain of validity of the adopted GMPEs (i.e., V_{S30} larger than 1200–1500 m.s^{-1}). In this case the GMPEs need to be adjusted for such hard-rock conditions in order to correctly represent the input motion for the site response analyses (see Sect. 4.3).

- choice of acceleration time histories when numerical simulations are foreseen. Selection of relevant time histories is an important step, especially for nonlinear analyses and represents an important task by itself; this aspect is covered in Sect. 6.7. Within the framework of SIGMA several different strategies have been used to choose the accelerograms based on the UHS: scaled natural records over the whole frequency range or over two different frequency ranges (2 sets of accelerograms) and spectrally-matched accelerograms.
- geometric definition of the soil profile: 1D (i.e., horizontally layered strata), 2D (i.e., alluvium valley or topographic ridge), or fully 3D.
- rheological characterization of the soil layers: linear viscoelastic, viscoelastic linear equivalent or fully non-linear behaviour.
- definition of the incident wave field: plane wave with vertical or oblique incidence, or surface wave (typically corresponding to a remote source); for a potential source close to the site, it is preferable to include the source in the computational model; this can be either a point source or an extended source, the latter one being more realistic for a source-site distance smaller or comparable to the fault size, and being preferable for earthquakes with magnitude larger than 6.
- choice of a (or several) software package to make the calculations; the most commonly used codes are listed in Sect. 6.3.2.

In SIGMA, almost all options were tested but most of the calculations were restricted to vertical incident plane waves.

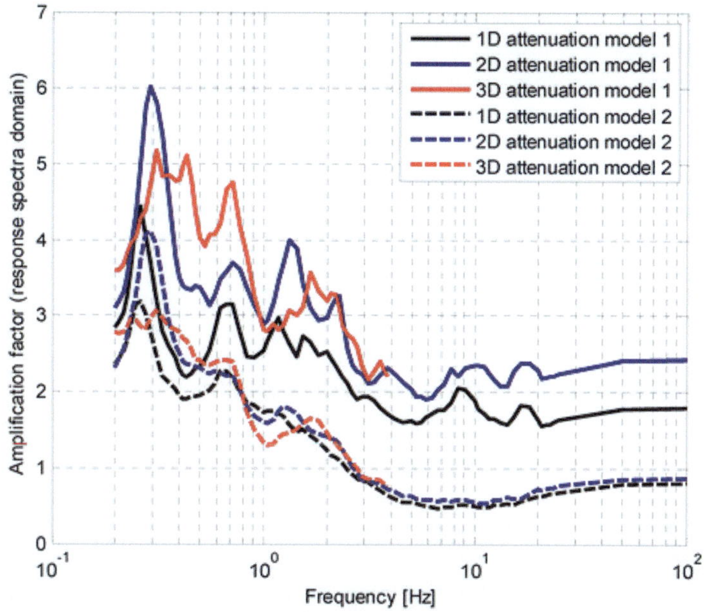

Fig. 5.12 Comparison between different attenuation models for 1D – 2D – 3D linear computations for the Grenoble test-site. *Solid line*: results with "model 1" taking into account standard scaling to define Q factors (e.g. $Q_S = V_S/10$). *Dashed lines*: results with "model 2", using the damping factor computed with 1D non-linear computations

5.3.1 Linear Numerical Analyses

From a rheological point of view, the linear viscoelastic assumption is the simplest constitutive model that can be used: it simply requires the definition of the dilatational V_P and shear V_S wave velocities and of attenuation (quality factor) for each wave, Q_P and Q_S.[1] While appropriate techniques exist for measuring the wave velocities (see Sect. 5.1), there are presently no established ones for measuring the attenuation; a typical, very crude rule of thumb is to use $Q = V_S/10$ (m.s^{-1}). Furthermore, no distinction is made between Q_P and Q_S. Sensitivity analyses carried out during the project have shown that the results (ground surface motions) are very sensitive to the choice of Q. Consequently, this parameter significantly increases the epistemic uncertainty of the analyses. To illustrate this point, Fig. 5.12 presents for the Grenoble site the surface response spectra calculated with two assumptions for attenuation.

The main advantage of linear constitutive models is their simplicity that theoretically allows considering 2D and, possibly 3D, geometries. However, in practice, 3D calculations are not feasible for frequencies higher than about 4 Hz because of the

[1] In geotechnical earthquake engineering the equivalent damping ratio, β, is more commonly used; it is related to the attenuation by $2\beta = Q^{-1}$

restrictions posed on the element mesh size ($h \leq \lambda/10$, with λ the wavelength) and above all because it is impossible to characterize the soil medium at such a small scale over large areas. Except for low amplitude rock motions, the linear viscoelastic analyses are not recommended as it is recognized that soils may exhibit nonlinear behaviour from small strains, although some counterexamples exist [see for instance Fig. 5.10 which documents linear behaviour up to PSV = 20 cm.s^{-1}, or the Mirandola site studied in Faccioli et al. (2015)]. Linear calculations may, however, be of value for calibrating the velocity profiles with the results of small recorded earthquakes or validating the computational codes in the linear range.

5.3.2 Equivalent Linear Numerical Analyses

Given the complexity of the nonlinear behaviour of soils, equivalent linear models represent a good compromise between engineering practice and scientific knowledge; they presently constitute the state of practice in site response analyses. Many models and codes have been developed and are currently used for such simulations (see Sect. 6.3.2). However, these models have some limitations: it is generally accepted that they are valid for shear strains smaller than 0.1–0.3%. Since the upper limit for the shear strain depends on the soil plasticity index and confining pressure (depth), it is better to relate it, rather than to an absolute value, to a reference shear strain γ_r defined as $\tau_{max} = \gamma_r$. $G_{max;}$ a value $\gamma_r = 2$ is suggested. Despite their relative simplicity and the limited number of parameters required (wave-velocity profiles, variation of the properties with shear strain), the equivalent linear models are also subject to large uncertainties. As noted in Sect. 5.1.4, the variations of properties can only be measured in the laboratory on undisturbed samples, and are not always obtained under the relevant stresses, especially for large depths. In view of these difficulties, often the curves $G/G_{max} = f(\gamma)$ and $\beta = g(\gamma)$ are chosen from published results in the literature. This uncertainty in the definition of the nonlinear properties is strongly reflected in uncertainties in the calculations. One such example is presented in Figs. 5.13 and 5.14, obtained for the Casaglia site (Deliverable D3-96) with different sets of published curves.

Equivalent linear analyses have been implemented both by the Italian (Casaglia site) and the French teams (Grenoble and EuroseisTest) using 1D calculations. Results from the Casaglia site are presented in Fig. 5.8 introduced previously. The two red solid curves correspond to two different strategies for the selection of the input motions: either natural, moderately scaled, or spectrally-matched records. The correction of the input signals for spectral matching does not affect values beyond about 0.2 s, where a close agreement with the NTC2008 norms (Italian code spectrum for Class C soil) is attained. For T < 0.2 s, on the other hand, the input correction slightly affects the results; in particular, the PGA value of the spectrum. More significant, however, is that the use of both SAFs leads to decreasing the values of the site specific spectra below the bedrock values at periods 0 < T < 0.2 s, mostly

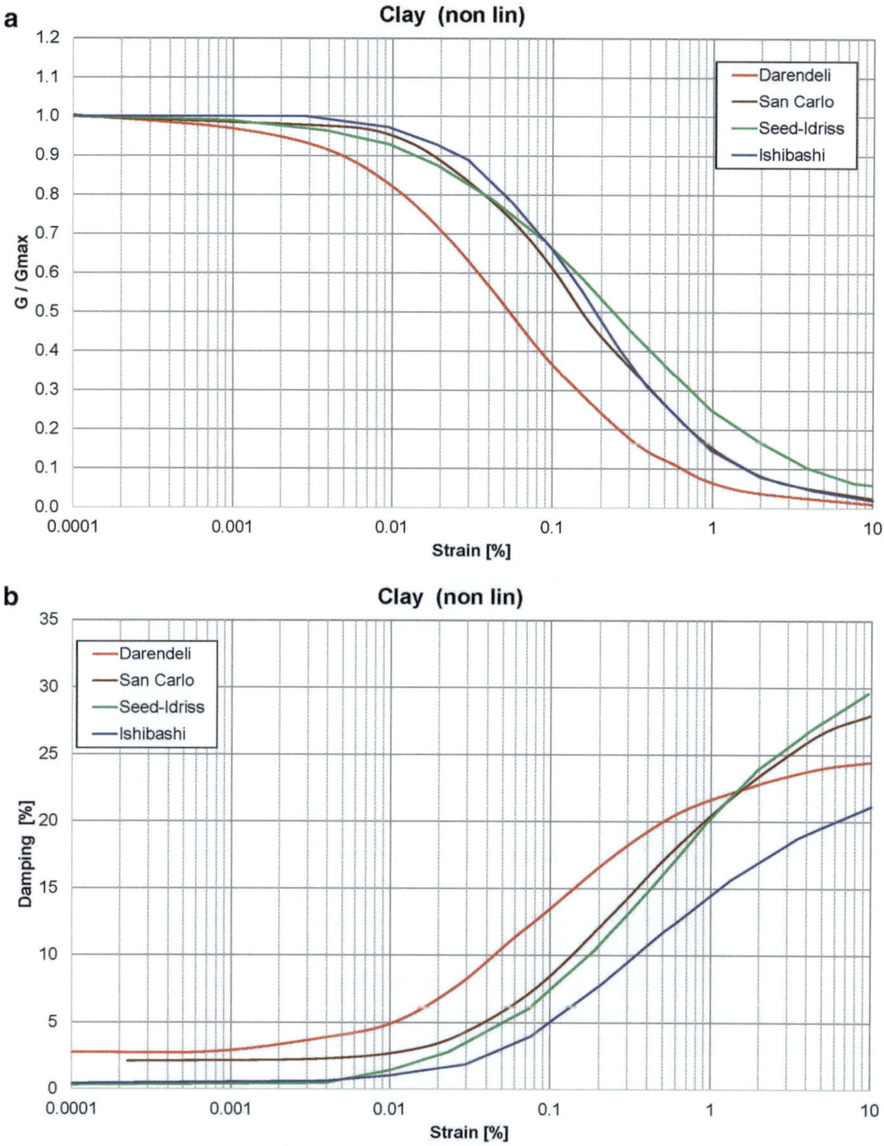

Fig. 5.13 Modulus reduction and damping curves for clay soils data at 9.6 m depth (from D3-96)

due to nonlinear effects. Also worth noting is that the non-ergodic PSHA spectrum from the GMPE (green curve) is not far from the "rock × SAF" spectra. On the whole, the Casaglia example strongly indicates that the soil surface spectra may decrease significantly when using results from local site analyses (such as SAFs or single site coefficients and sigmas) instead of the site coefficient of the GMPEs.

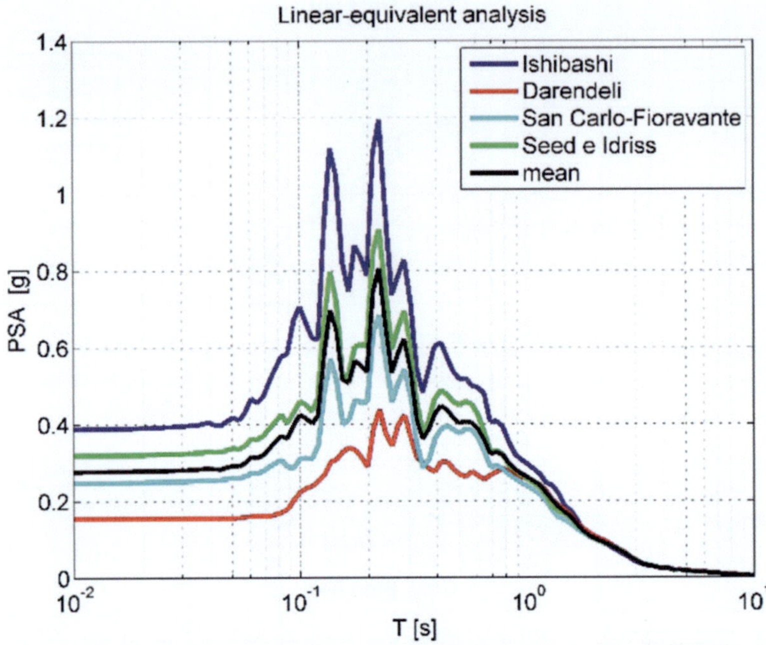

Fig. 5.14 Five percent damped response spectra obtained by 1D propagation analyses performed using the soil degradation curves summarized in Fig. 5.13 (From D3-96)

This is caused by the notable deamplification indicated by negative $\delta S2S$ values and does not necessarily hold true for all sites.

Results obtained by the French team for the Grenoble site shown in Fig. 5.15 are consistent with those from Casaglia. The SAF applied to the rock spectrum are derived from 1D linear, equivalent linear analyses after the bedrock motion has been corrected for hard rock conditions ($V_S = 3500$ m.s^{-1}). The surface spectra do exhibit the same marked reduction from the spectrum calculated with the generic GMPE (black solid line) when equivalent linear models are used and predict a much larger amplification when using a linear model. Again, the set of time histories selected by the Italian team (IT) or by the French team (FR) does not appear to have a strong impact on the results.

Additional calculations were carried out for 2D and 3D geometries and linear soil behaviour. The results, not presented herein (see Deliverable D3-152), show that the aggravation factor, which intends to reflect the impact of the sub-surface topography by correcting the surface spectrum leads to a surface spectrum that is much higher (except at high frequencies >20 Hz) than the spectrum directly calculated from 3D, and even 2D, linear analyses; use of aggravation factors therefore appears to be very conservative. Note, however, that due to inherent limitations of 3D calculations (see Sect. 5.3.1) they could not have been carried out above 4 Hz and the 3D results at larger frequencies have been extrapolated based on the SAFs

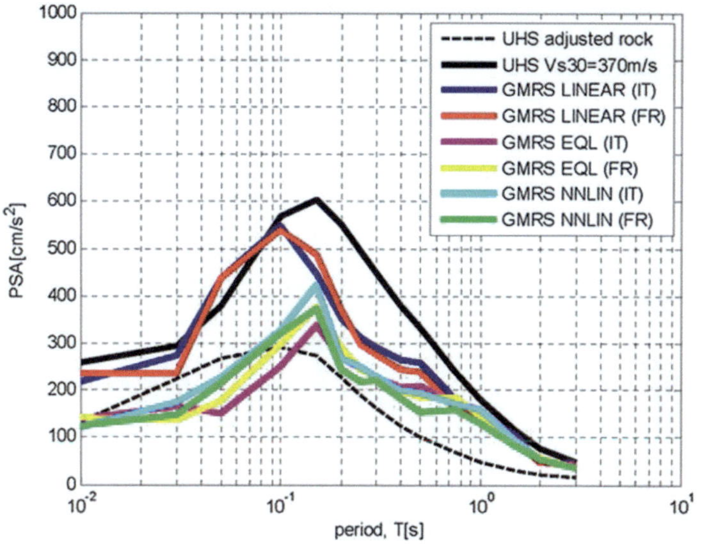

Fig. 5.15 Response spectra computed for the Grenoble test-site with different approaches (all target levels are for a return period of 10,000 years). Comparison between generic mean UHS (*black*) and site-specific mean GMRS obtained using linear (*blue, red*), equivalent-linear (*magenta*) and nonlinear (*light blue, green*) site amplification factors and Vs-kappa adjustment. The mean UHS for bedrock conditions is also shown by the *dashed black curve*

of the 2D calculations; this may explain the somewhat inconsistent results, which are nevertheless also predicted by 2D linear calculations.

5.3.3 Nonlinear Numerical Analyses

When the shear strain exceeds the amplitudes previously indicated (i.e., above 0.3–0.5%, or better $\gamma > 2\gamma_r$) or, in other words, for very soft soils and/or very severe loading, a complete nonlinear modelling would theoretically be required, with an appropriate constitutive relationship and its associated soil parameters. These models oscillate between two poles: relatively simple constitutive models with few parameters, which can hardly reproduce all possible loading/unloading paths, and more complex models with many parameters (sometimes exceeding ten), which may succeed in describing all possible paths, but whose determination remain largely beyond experimental capabilities. The validation of nonlinear approaches is a major issue, which has not, up to now, received satisfactory scientific answers; the Prenolin international benchmark was launched to address these issues. Twenty three world wide teams participated in the benchmark (D3-114 and D3-149) with the major objective of assessing the actual epistemic uncertainties associated with nonlinear calculations, emphasizing those associated with the software (constitutive

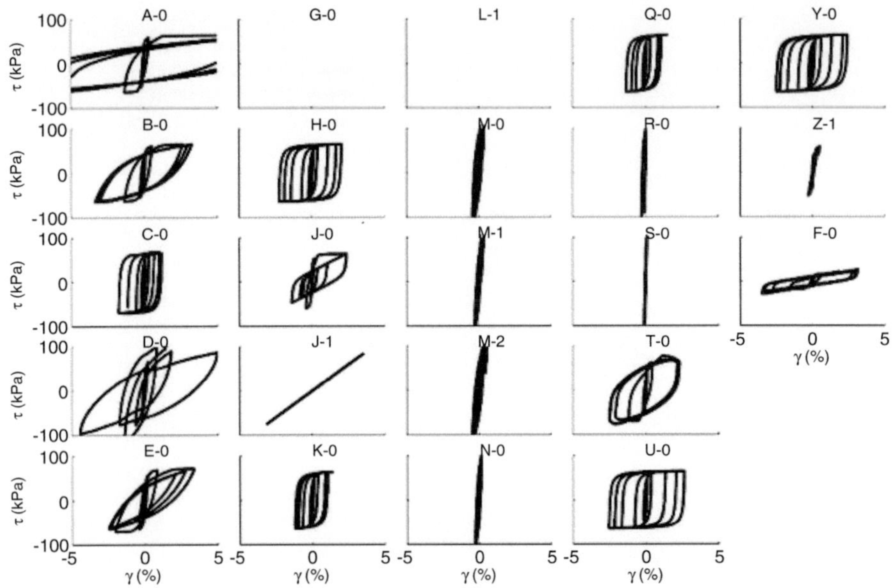

Fig. 5.16 Cyclic stress-strain loops for a soil element with shear strength 65 kPa subjected to a sinusoidal input seismic motion of 10 s. Predictions produced by the different constitutive models

models and numerical scheme), those associated with the translation of raw field and laboratory data into design values for the nonlinear model parameters, and the actual deviation of numerical predictions from observations (bias and scatter). Two sites have been chosen in Japan and extensive 1D nonlinear calculations have been performed. One of the main lessons from this project is the importance of the choice and calibration of the constitutive model; this implies numerical testing of the model over the whole strain range and comparison with laboratory tests. Figure 5.16 presents results of the numerical tests of the constitutive models used in Prenolin deemed to represent the cyclic behaviour of the same soil element for which the same mechanical characteristics were provided to all participants.

Many results have been produced by this benchmark; their post-processing is still ongoing and cannot be summarized here. An illustration of comparisons between computed nonlinear site response and that observed is depicted in Fig. 5.17 for four input motions; the weakest input motion is #9 and the two strongest ones are #1 and #2.

Preliminary conclusions from the benchmark indicate that:

- a very careful characterization of the soil profile is needed to achieve adequate modelling with a combination of both laboratory and site measurements implemented with different techniques (invasive *vs.* non-invasive);
- use of several non-linear codes operated by different teams is desirable to assess the epistemic uncertainty associated with the constitutive models; and

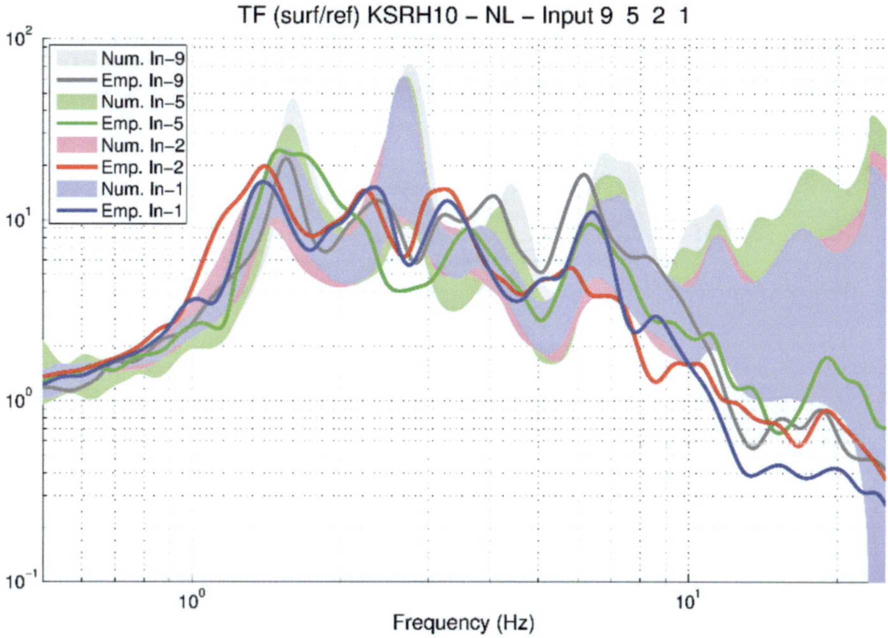

Fig. 5.17 Comparison between the predicted transfer function at KiK-net site KSRH10 (16–84 percentiles, *colour shaded areas*) and the observations (*solid lines with corresponding colours*), for input motions 9, 5, 2 and 1

- despite the large differences in the soil constitutive models and numerical schemes, the epistemic uncertainty of the site-specific nonlinear calculations is smaller than the value of the within-event variability of GMPEs for the site.

Nonlinear numerical analyses were also tested for the three sites: Casaglia, Grenoble and EuroseisTest. Only one nonlinear constitutive model was retained for each case and the results were compared to those of linear and equivalent linear calculations. Results are presented only for one site (Grenoble) in Fig. 5.18; however, the same conclusions were reached for the two other sites (see Fig. 5.8 for Casaglia):

- Linear calculations produce the highest amplification; as discussed previously, this is due in a large part to the inaccurate definition of damping but, nevertheless, some sites may exhibit a more linear response than others;
- Linear equivalent and nonlinear models tend to predict SAFs of the same order of magnitude except for large amplitude motions (top diagram in Fig. 5.18 corresponding to a large return period); and
- SAFs are not very sensitive to the set of accelerograms used for the calculations:

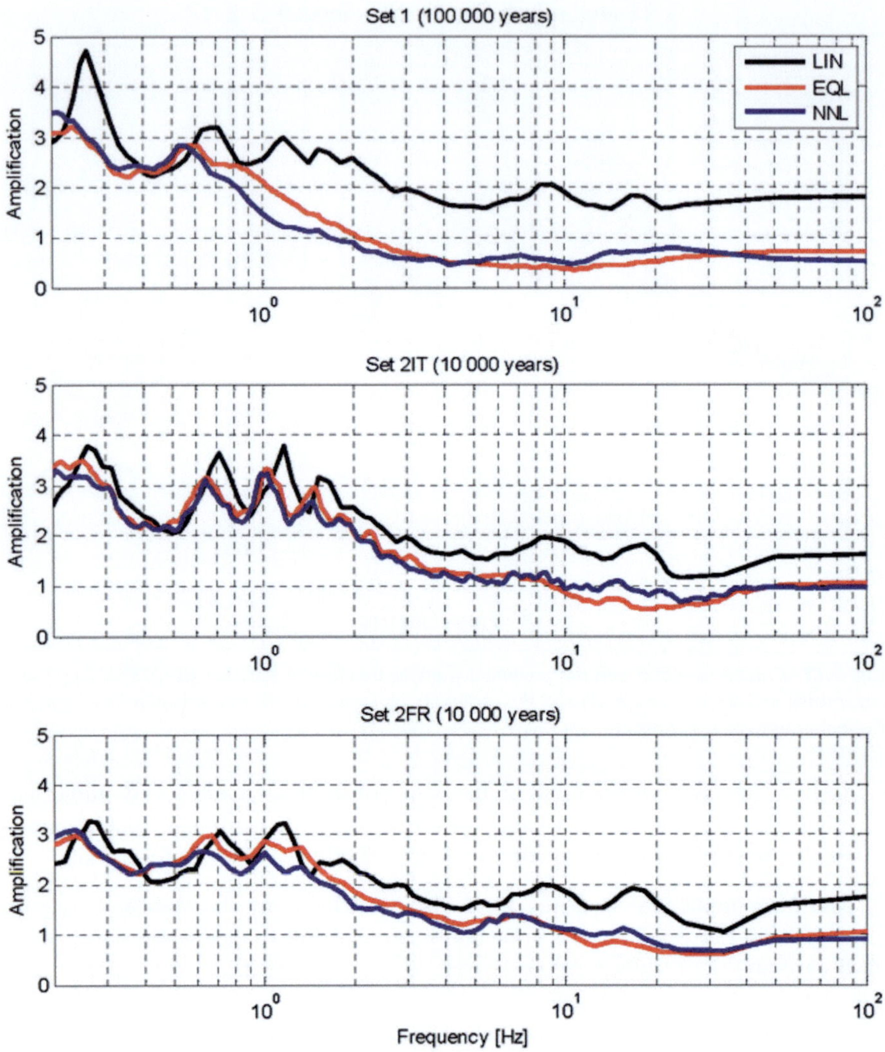

Fig. 5.18 Spectral amplification factor computed in 1D analyses for the Grenoble test-site, with accelerometers from set 1 (*top*), set 2IT (*middle*) and set 2FR (*bottom*)

- Set 2FR: accelerograms representative of a smaller motion corresponding to a return period of 10,000 years and a PGA of 0.15 g, proposed by the French team; and
- Set 2IT: spectrally matched accelerograms proposed by the Italian team for the same target as above.

In terms of ground response surface spectra Fig. 5.15 shows that, as noted previously for the Casaglia site, the estimated soil surface spectra decrease significantly when using results from local site analyses (such as SAFs) instead of the site coefficient of the GMPEs. Furthermore, except for very large input motions, complex nonlinear analyses are not really warranted and equivalent linear analyses appear to be accurate enough for engineering purposes.

5.4 Treatment of Uncertainties

The general framework for the treatment of uncertainties has been presented in Sect. 4.4 for the rock hazard. This paragraph explains how the uncertainties for site hazard were handled in the SIGMA project.

5.4.1 *Fully Probabilistic Generic Site Approach (FpG)*

For site hazard, the same scheme applies when the FpG approach is implemented with the site term $\delta S2S$ and its site-to-site variability ϕ_{S2S} related to the soil site. The total single-station standard deviation at an individual site is then given by the last line of Table 4.3. As already cited in Sect. 5.2.1 the methodology has been applied to 12 sites in the Po Plain by Faccioli et al. (2015) to calculate the standard deviation of $\delta S2S$ and $\phi_{ss,s}$, the event corrected single-station standard deviation at individual sites. The epistemic uncertainty on $\phi_{ss,s}$, was estimated from its standard deviation across many stations (in this case, the Po Plain 12-station set), thereby implicitly assuming ergodicity in the variance (not in the mean). Such standard deviation (std ($\phi_{ss,s}$)) was found equal to 0.08, and used to define upper and lower bound estimates of $\sigma_{ss,s}$ according to:

$$\sigma_{ss,s} = \sqrt{\left(\phi_{ss,s} \mp std\left(\phi_{ss,s}\right)\right)^2 + \tau^2} \tag{5.2}$$

τ is the inter-event variability component of the considered GMPE (ITA13, see deliverable D2-72), and its uncertainty was neglected.

The foregoing variability estimates are illustrated in Fig. 5.19, which also shows that the ITA13 GMPE sigma represents a kind of average $\sigma_{ss,s}$ of the three sites. Introducing the variability bounds through $\phi_{ss,s}$ is intended to accommodate, at least in part, the uncertainty caused by multipathing, which is only partially accounted for in the available data because of the predominance of the 2012 Emilia sequence records both at the study sites and in the ITA13 GMPE (but see, on this aspect, footnote 6 in Chap. 4).

The previous median (Eq. 5.1) and sigma (Eq. 5.2) formulations have been combined in a simple logic tree for seismic hazard estimation. Comparison of predictions with data from actual records (Reggio Emilia 1992 and Emilia 2012 earthquakes) shows that spectra are mostly within the spread spanned by the logic-

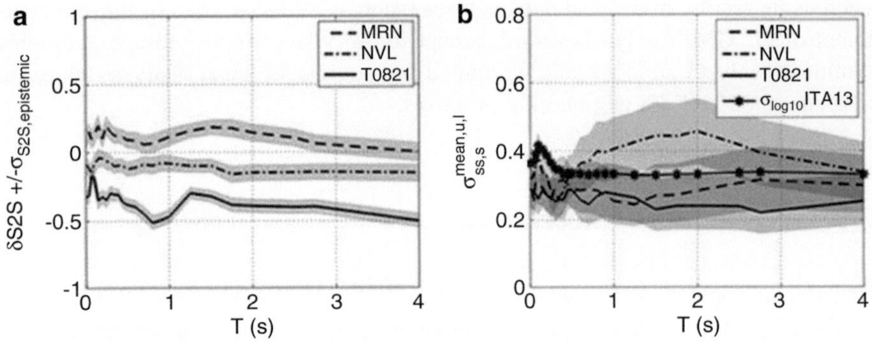

Fig. 5.19 (**a**) Site terms δS2S for the three study sites with $\pm\sigma_{S2S;epistemic}$ bands (T0821 coincides with CAS). (**b**) Total single-site sigma $\sigma_{ss;s}$ for the same sites (*lines*), with variability bands enclosed within the upper and lower limits estimated from Eq. (5.2); the ITA13 GMPE standard deviation ($\sigma_{\log 10}$ ITA13) is also shown in this plot. The variability estimates are in both cases derived from a 12-site Po Plain data subset

tree branches for a return period of 475 years, and the spectral shapes are reasonably similar. The code spectra are also consistent with the UHS at 475 years.

5.4.2 Hybrid Site Specific Approach (HyS)

Two main sources of uncertainties are involved in the hybrid-approach calculation of site-specific hazard, each one containing an aleatory and an epistemic component:

- uncertainties in the rock (or hard-rock) hazard. Most of the ground motion uncertainties are accounted for in the standard deviation of the GMPEs and in host-to-target adjustment, if applied.
- uncertainties in the site amplification function.

Apart from the aleatory uncertainty related to nonlinear soil behaviour, most of the other sources of uncertainty are already, to a large extent, included in the aleatory variability of the rock GMPEs. For this reason, except for cases where the variability of nonlinear behaviour is significant, *aleatory* variability in site response must not be included to avoid double counting of aleatory uncertainties in rock and site hazard.

For a given target spectrum on rock, the sources of uncertainties in site response evaluation can be identified as related to: (a) selection of input motions, (b) dynamic properties of soil profile, (c) selection of the method of analysis for site response evaluation, and (d) modelling of soil behaviour. To avoid double counting of aleatory uncertainties, only epistemic contributions will be referred to in the following, so that the small scale random variability of soil properties is disregarded. As a matter of fact, the impact of the small scale random variability in the mechanical properties of the propagation path already can be considered as entering into the ϕ_{ss} uncertainty component of the partially non-ergodic analysis, by which the bedrock UHS were determined.

The impact of the various sources of epistemic uncertainties in site response has been illustrated in SIGMA with the studies reported in Faccioli et al. (2015) on one site in the Po plain, i.e. the accelerometer station Mirandola (MRN).

5.4.2.1 Uncertainty Related to Selection of Input Motions

It is shown that the uncertainty linked to the choice of input motions is considerably reduced, to become almost negligible in the total uncertainty, when broad-band spectral matching of the records to the target spectrum is achieved (see Sect. 6.7.1).

5.4.2.2 Uncertainty in V_S Profiles

As pointed out in Sect. 5.1, this uncertainty can be reduced by combining experimental measurements, measurements of velocity profiles by different techniques including invasive and non-invasive techniques. For the Mirandola site studied by Faccioli et al. (2015), the standard deviation of the site response due to alternative V_S profiles, $\sigma_{Vs\ log10}$, was found to be of the order of 0.05–0.08 (Fig. 5.20).

5.4.2.3 Uncertainty in Soil Behaviour Modelling

The examples presented either for the Casaglia site (Fig. 5.14) and the Grenoble site (Fig. 5.18) have shown that the largest epistemic uncertainty resides in the choice of the constitutive model (linear, equivalent linear or nonlinear) and of the associated

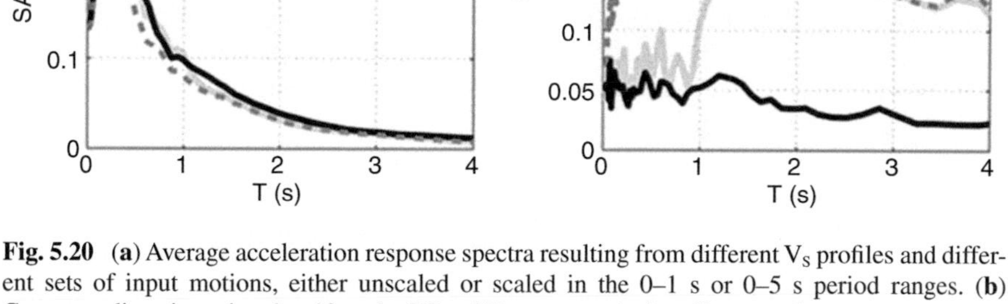

Fig. 5.20 (**a**) Average acceleration response spectra resulting from different V_S profiles and different sets of input motions, either unscaled or scaled in the 0–1 s or 0–5 s period ranges. (**b**) Corresponding sigma in a log 10 scale. RP = 475 years, equivalent-linear analyses

parameters (degradation curves of Fig. 5.13). Although the Italian team found that for the deep soil sites in the Po plain the linear model predicted the closest response to the recorded motions, this conclusion cannot be generalized to any situation: it obviously depends on the soil material, the intensity of the rock motion, the reliability of the soil parameters and so forth.

5.4.2.4 Combination of Epistemic Uncertainties

Faccioli et al. (2015) proposed that the resulting σ_{TOT} associated with the average site-specific response spectra for a given return period, including both the epistemic uncertainties of the site-response analysis and the total (aleatory + epistemic) uncertainties carried by the PSHA at exposed bedrock ($\sigma_{PSHA\ rock}$) be evaluated as

$$\sigma_{TOT} = \max_{T}\left(\sqrt{\sigma_{Vs}^2 + \sigma_{soil\_model}^2 + \sigma_{PSHA\_rock}^2}; \sigma_{Kik\ net}\right) \tag{5.3}$$

where $\sigma_{Kik\text{-}net}$ is the standard deviation of SAFs of 21 deep soil sites considered from the KiK-net (Sect. 5.2.1). The latter values of σ, which are in general related to the combination of a wider set of seismic site amplification factors, rather than 1D effects alone, can reasonably be considered as a lower bound for evaluations of site-specific variability of results.

5.5 Lessons Learned from the SIGMA Project

The major lessons in site response analyses learned from the SIGMA project have been highlighted in the present chapter. The following bullet points are simply a summary of the most important ones:

- Consideration of site response in SHA requires more or less detailed characterization of the soils, depending on the level of sophistication of the analyses; this characterization should be obtained from tests (field and/or laboratory tests) and can be advantageously complemented with site instrumentation. The minimum required information should include:
  - Description of the geometry of softer layers, with respect to underlying bedrock, and especially, in cases where topographic effects are important, determination of the aspect ratio (ratio between height and width of a basin);
  - position of the considered site with respect to the basin border;
  - V_S and V_P profiles; and
  - shear-wave velocity contrast between basin (or more generally the soft layers) and bedrock.
- Whenever possible the single-station sigma approach should be used with a view to decreasing the level of uncertainty in the ground surface response.

- The Hybrid site specific approach (HyS) is the most versatile approach and represents a good compromise between the simple fully probabilistic generic approach (FpG), which tends to overestimate the ground surface response, and the fully probabilistic site-specific approach (FpS), which is very demanding and does not necessarily address all aspects of site response.
- In the HyS approach the main source of epistemic uncertainty resides in the soil behavior modelling; several models are available from the simple viscoelastic linear model to the fully nonlinear model. The most appropriate one is still a matter of debate and comparison with field records gives contradictory results. Furthermore, characterization of the constitutive model parameters is prone to large uncertainties. At the present stage, it seems that equivalent linear models present a sufficient degree of accuracy and recourse to more sophisticated nonlinear models is not warranted, except maybe at very long return periods.

Based on the experience gained during the SIGMA project, guidelines to consider site response in PSHA have been prepared as deliverable D3-152. The main objective of the guidelines is to propose a gradual approach based on the HpG, FpG and HyS methods for considering site response; the choice between the three approaches depends on the level of characterization of the input data and the estimated importance of site response.

5.6 Additional Topics in Ground Surface Hazard Assessment

The following topics, although important for seismic hazard assessment, have deliberately not been considered in SIGMA. Therefore, they are not discussed in detail and only general concepts are presented in the following paragraphs.

5.6.1 Vertical Ground Motion

The number of GMPEs for vertical motions is, by far, fewer than those for horizontal motions. Therefore, among the approaches listed in Table 5.1, the most commonly implemented are the HyG, HyS and FpS approaches. HyS and FpS are based on numerical analyses. HyG is usually implemented after having calculated the horizontal UHS at the ground surface (by any of the four approaches) and applying V/H ratios, i.e. ratios between the vertical and horizontal response spectra at the ground surface. Statistical frequency-dependent relationships have been developed for such ratios. In this approach, attention must be paid to large amplitude motions that may induce significant nonlinearities in the soil under horizontal loading. Under vertical loading, mainly caused by the vertical propagation of dilatational waves, nonlinearities are much more limited; it is, therefore, important that the V/H relationships reflect this characteristic, otherwise the vertical motion may be underestimated. Few such relationships exist; one can refer to Gülerce and Abrahamson, (2011).

Another approach for estimating the vertical ground motion may be based on physics: for shallow profiles, especially those composed of water-saturated materials, the wave lengths are much longer than the profile thickness because the soil is nearly incompressible; the dilatational wave velocity exceeds 1500 $m.s^{-1}$ (velocity of sound in water). It can, therefore, be argued that the vertical rock motion is barely amplified at the ground surface.

Finally, if the ground surface response spectrum is calculated from 1D site response analyses (HyS or FpS) attention must be paid to the soil characteristics. Usually the vertical motion is calculated independently of the horizontal one and the question arises of the proper soil characteristics to enter in the model, at least for medium-to-large-amplitude motions. The state of practice is to use the properties (V_S) retrieved from the horizontal motion calculation (the strain compatible characteristics in equivalent linear analyses) and to convert them into V_P. In saturated soils P waves travel through the water; the bulk modulus of the soil skeleton may be slightly affected by the induced shear strain but the overall bulk modulus, which is the sum of the soil skeleton bulk modulus and of the water bulk modulus, will be almost unaffected. In dry soils, the propagation of P waves is controlled by the skeleton properties:

$$\rho V_p^{\,2} = K + \frac{4}{3}G \tag{5.4}$$

G and K are influenced by the shear strain but the bulk modulus K to a much lesser extent than the shear modulus G. It seems, therefore, appropriate to calculate V_P for the vertical analyses assuming that K is strain independent, equal to its elastic value, and that G is equal to the strain compatible value calculated in the horizontal equivalent linear runs. Applying the same reduction to V_P as to V_S to account for nonlinear effects is certainly unrealistic; however, any reduction in-between the one proposed above, with no reduction on K, and the full reduction coming from V_S is debatable and contributes to the epistemic uncertainty. Obviously, if full nonlinear analyses are carried out for the horizontal motion, the vertical motion shall be input simultaneously to the horizontal one and the constitutive model will account for the (time) variations of both wave velocities.

5.6.2 *Maximum Ground Motion: Truncation*

Recent studies to assess very long-term seismic hazard in the United States and in Europe have brought the issue of upper limits on earthquake ground motions into the arena of problems requiring attention from the engineering seismological community. Few engineering projects are considered sufficiently critical to warrant the use of annual frequencies of exceedance so low that ground-motion estimates may become unphysical if limiting factors are not considered, but for nuclear waste repositories, for example, the issue is of great importance. The definition of upper bounds on earthquake ground motions also presents an exciting challenge for researchers in the area of seismic hazard assessment (Bommer et al. 2004).

The maximum ground motions that can be experienced at the ground surface are controlled by three factors: the most intense seismic radiation that can emanate from the source of the earthquake; the interaction of radiation from different parts of the source and from different travel paths; and the limits on the strongest motion that can be transmitted to the surface by shallow geological materials. As seismic waves propagate to the Earth's surface, other factors act to limit the maximum amplitude of the motion. These factors are associated with the failure of surface materials, which are usually weaker than the underlying rock, under the loading conditions generated by the passage of seismic radiation. The principle is similar to that of a fuse: once failure is reached at a given depth within the soil profile, the high frequency components of the incident motion are filtered and accelerations larger than those reached at the failure stage cannot be transmitted to the upper strata; however, this is counterbalanced by larger displacements.

The first and most obvious tool for exploring upper bounds on ground motions is the ever-increasing databank of strong-motion accelerograms. Another approach is to define the maximum rock motion, if any, and to calculate, based on the strength of the overlying soil materials, the maximum motion that can be transmitted either from simplified analytical models (Pecker 2005) or from nonlinear site response analyses with increasing amplitudes at the bedrock. The simple procedure of truncating at a specified number of logarithmic standard deviations above the logarithmic mean of the GMPE is unlikely to be an adequate solution (Bommer et al. 2004), see also Sect. 4.4.1.

As a conclusion of this brief discussion, the issue of maximum ground motions is still an open question requiring further attention.

References

Bazzurro P, Cornell CA (2004a) Ground-motion amplification in nonlinear soil sites with uncertain properties. Bull Seismol Soc Am 94(6):2090–2109

Bazzurro P, Cornell CA (2004b) Nonlinear soil-site effects in probabilistic seismic-hazard analysis. Bull Seismol Soc Am 94(6):2110–2123

Bommer JJ, Abrahamson N, Strasser FO, Pecker A, Bard PY, Bungum H, Cotton F, Fäh D, Sabetta F, Scherbaum F, Studer J (2004) The challenge of defining upper bounds on earthquake ground motions. Seismol Res Lett 75(1):82–95

Cadet H, Bard P-Y, Rodriguez-Marek A (2012a) Site effect assessment using KiK-net data: Part 1. A simple correction procedure for surface/downhole spectral ratios. Bull Earthq Eng 10(2):421–448

Cadet H, Bard P-Y, Duval AM, Bertrand E (2012b) Site effect assessment using KiK-net data: Part 2: Site amplification prediction equation based on f_0 and VSZ. Bull Earthq Eng 10(2):451–489

Chavez GF, Faccioli E (2000) Complex site effects and building codes: making the leap. J Seismol 4(1):23–40

Cramer CH (2003) Site seismic-hazard analysis that is completely probabilistic. Bull Seismol Soc Am 93(4):1841–1846

Faccioli E, Paolucci R, Vanini M (2015) Evaluation of probabilistic site-specific seismic hazard methods and associated uncertainties, with applications in the Po Plain, Northern Italy. Bull Seismol Soc Am 105(5):2787–2807

Foti S, Lai CG, Rix GJ, Strobbia C (2014) Surface wave methods for near-surface site characterization. CRC Press, Boca Raton, p 487, ISBN:9780415678766

Garofalo F, Foti S, Hollender F, Bard PY, Cornou C, Cox BR, Dechamp A, Ohrnberger M, Perron V, Sicilia D, Teague D, Vergniault C (2016) InterPACIFIC project: comparison of invasive and non-invasive methods for seismic site characterization, Part II: Inter-comparison between surface-wave and borehole methods. Soil Dyn Earthq Eng 82:241–254

Gülerce Z, Abrahamson NA (2011) Site-specific spectra for vertical ground motion. Earthquake Spectra 27(4):1023–1047

Hardin BO, Drnevich VP (1972) Shear modulus and damping in soils: design equations and curves. J Soil Mech Found Div ASCE 98(SM6):667–692

Ishihara K (1996) Soil behaviour in earthquake geotechnics, Oxford Engineering Science Series. Oxford University Press, Oxford, p 342

Nakamura Y (1989) A method for dynamic characteristics estimation of subsurface using microtremor on the ground surface, Quarterly Report Railway Technical Research Institute, 30–1, 25–33, Tokyo, Japan

Ordaz M, Singh SK, Arciniega A (1994) Bayesian attenuation regressions: an application to Mexico City. Geophys J Int 117:335–344

Pecker A (2005) Maximum ground surface motions in probabilistic seismic hazard analyses. J Earthq Eng 9(4):1–25

Perez A, Jaimes MA, Ordaz M (2009) Spectral attenuation relations at soft sites based on existing attenuation relations for rock sites. J Earthq Eng 13(2):236–251

Régnier J, Cadet H, Bonilla LF, Bertrand E, Semblat JF (2013) Assessing nonlinear behaviour of soils in seismic site response: statistical analysis on KiK-net strong-motion data. Bull Seismol Soc Am 103(3):1750–1770

Renault P (2009) PEGASOS/PRP Overview. Joint ICTP/IAEA advanced workshop on earthquake engineering for nuclear facilities. Abdus Salam International Center for Theoretical Physics, Trieste, Italy

Rodriguez-Marek A, Rathje E, Bommer J, Scherbaum F, Stafford PJ (2014) Application of single-station sigma and site-response characterization in a probabilistic seismic-hazard analysis for a new nuclear site. Bull Seism Soc Am 104(4):1601–1619

SESAME (2004) European research project WP12, Guidelines for the implementation of the H/V spectral ratio technique on ambient vibrations – measurements, processing and interpretation – Deliverable D23.12. European Commission – Research General Directorate Project No. EVG1-CT-2000-00026 SESAME

Chapter 6
Seismic Hazard Computation

This section deals with the seismic hazard computation process. It should be noted that, in cases where the seismic hazard to be calculated includes site-specific soil amplifications, this process may include two steps: hazard calculation for reference rock conditions and subsequent site response analysis to combine the local soil response with the rock hazard (see Sect. 5.2).

6.1 Basic Requirements

The following basic requirements apply to the seismic hazard computation.

- In large projects, the teams in charge of the development of the Ground Motion (GM) models (SIGMA WP2) and of the Seismic Source (SS) models (SIGMA WP1 and WP4) are often different from the Hazard Calculation Team (HCT). For this reason, it is vital for the project's success to establish good communication and interaction between the HCT and the model developers (Seismic Source Characterization - SSC, and Ground Motion Characterization - GMC, teams) to ensure full agreement between the model finally implemented in the PSHA software and the models developed by the SSC and GMC teams.
- It is desirable, and of great benefit for the project, that the SSC and GMC teams produce Hazard Input Documents (HIDs) to be provided to the HCT (Ameri 2015). The HID provides a list of input parameters required for the PSHA calculation within the project. The final HID should provide sufficient information to the HCT to build the input to the SSC and GMC logic tree in the hazard code for PSHA calculations, and to avoid any non-unique interpretation. Several interactions between the GMC-SSC teams and the HCT may be required in order to refine the HID and to verify that consensus is reached on all the necessary input parameters to the software.

A. Pecker et al., *An Overview of the SIGMA Research Project*, Geotechnical, Geological and Earthquake Engineering 42, DOI 10.1007/978-3-319-58154-5_6

- Software verification and validation (V&V) documentation: in recent years, the requirements related to software qualification were formally included in international guides (e.g. IAEA, SSG-9, USNRC RG 1.208). Thus, it is crucial to develop verification and validation plans to guarantee the users and end-users that the calculations performed with the software are reliable. Such Quality Assurance (QA) documents can also be available from previous studies for particular software (e.g., Thomas et al. 2010). It is, in any case, good practice to identify the capabilities of the selected software and the potential limitations in the application to the current project. Such limitations shall be identified early in the project to allow adopting another software if necessary and to verify that the software is flexible enough for integration of new calculation methods or for reducing the calculation time. If documentation on the software V&V is not available or judged insufficient, simple tests should be performed in order to verify the results for typical applications of concern for the project.

6.2 Interfaces and Boundary Conditions

The different steps involved in a PSHA (SSC, GMC and SRC) are often perceived as separate tasks that are finally combined into a unique logic tree at the stage of the hazard calculations. However, this perspective should be avoided as a number of obvious interfaces exist between all these tasks, and because the true inputs from SSC, GMC and SRC should be correctly reflected in the hazard calculation model. These interface issues are addressed in Sect. 2.6 and Chap. 7.

6.3 Software Packages

6.3.1 PSHA Software

Several PSHA software packages have been developed in the last decades. Some of them are licensed products and others are completely open-source. The choice of the PSHA software is not a trivial issue and it should be oriented on the basis of the specific needs of the project. Moreover, with such a large number of available software, the issues related to the reliability and accuracy of hazard results and to the Quality Assurance protocol become of great importance (e.g., Bommer et al. 2013). Another factor which might strongly influence the choice of a PSHA software is the acceptance by an independent reviewer or a regulatory body. Some of the available codes have an already accepted nuclear QA programme (e.g. by US NRC), while some others are mainly used in the academic environment, without appropriate V&V documentation.

<table>
<tr><td colspan="2">Features</td><td>CRISIS</td><td>EQRM</td><td>FRISK 88M</td><td>MoCaHAZ</td><td>MRS</td><td>NSHMP</td><td>OHAZ</td><td>OpenSHA</td><td>SEISRISK</td><td>SEISHAZ</td></tr>
<tr><td colspan="2">PSHA approach</td><td>classic</td><td>Monte Carlo</td><td>classic</td><td>Monte Carlo</td><td>classic</td><td>classic</td><td>classic</td><td>classic</td><td>classic</td><td>classic</td></tr>
<tr><td colspan="12">Seismic Source Types</td></tr>
<tr><td colspan="2">Area</td><td colspan="5">Yes</td><td>No</td><td colspan="3">Yes</td><td>No</td></tr>
<tr><td>Fault</td><td>Line [2D]</td><td>Yes</td><td rowspan="2">Limited</td><td rowspan="2">Yes</td><td rowspan="2">No</td><td rowspan="2">No</td><td rowspan="2">Yes</td><td>Yes</td><td rowspan="3">Yes</td><td>Yes</td><td rowspan="3">Yes</td></tr>
<tr><td></td><td>Volume [3D]</td><td>Yes</td><td>No</td><td rowspan="2">No</td></tr>
<tr><td colspan="2">Point</td><td>Yes</td><td>No</td><td>Yes</td><td>Yes</td><td>No</td><td>Yes</td><td>Yes</td></tr>
<tr><td colspan="2">Allow to assign a depth distribution to each source</td><td colspan="8">Yes</td><td>No</td><td>Yes</td></tr>
<tr><td colspan="2">Allow to assign a style-of faulting to each source</td><td>Limited</td><td colspan="2">Yes</td><td>No</td><td colspan="4">Yes</td><td>No</td><td>Yes</td></tr>
<tr><td colspan="2">Rupture length & width Modelling</td><td></td><td>Yes</td><td></td><td>No</td><td colspan="6">Yes</td></tr>
<tr><td colspan="12">Magnitude Frequency Distribution [MFD]</td></tr>
<tr><td colspan="2">Gutenberg-Richter</td><td colspan="10">Yes</td></tr>
<tr><td colspan="2">Gaussian</td><td>Yes</td><td>No</td><td>Yes</td><td>No</td><td>No</td><td>Yes</td><td>No</td><td>Yes</td><td>No</td><td>Yes</td></tr>
<tr><td colspan="2">Customize MFD intervals</td><td>No</td><td>Yes</td><td>No</td><td>No</td><td>No</td><td>Yes</td><td>No</td><td>Yes</td><td>Yes</td><td>No</td></tr>
<tr><td colspan="12">Ground Motion Prediction Equations [GMPE]</td></tr>
<tr><td colspan="2">Built-in</td><td colspan="8">Yes</td><td>No</td><td>Yes</td></tr>
<tr><td colspan="2">User Defined</td><td>Yes</td><td>No</td><td>Limited</td><td>Yes</td><td>No</td><td>No</td><td>Limited</td><td>No</td><td>Yes</td><td>No</td></tr>
<tr><td colspan="2">Allow to assign different GMPEs per seismic source type</td><td colspan="3">Yes</td><td>No</td><td colspan="4">Yes</td><td>No</td><td>Yes</td></tr>
<tr><td rowspan="2">Truncation</td><td>Max Ground Motion Value</td><td colspan="3">Yes</td><td colspan="4">No</td><td colspan="2">Yes</td><td>No</td></tr>
<tr><td>No. of Sigma</td><td colspan="10">Yes</td></tr>
<tr><td colspan="2">Variability</td><td colspan="10">Yes</td></tr>
<tr><td colspan="2">Customizable units</td><td colspan="4">Yes</td><td>No</td><td>Yes</td><td></td><td></td><td></td><td></td></tr>
<tr><td colspan="12">LOGIC TREE</td></tr>
<tr><td colspan="2">Allows to define a logic tree</td><td colspan="3">Yes</td><td>No</td><td>No</td><td>Yes</td><td>No</td><td>Yes</td><td>No</td><td>Yes</td></tr>
<tr><td colspan="12">OUTPUT</td></tr>
<tr><td colspan="2">Hazard Curves</td><td colspan="6">Yes</td><td>No</td><td colspan="3">Yes</td></tr>
<tr><td colspan="2">Hazard Maps</td><td colspan="4">Yes</td><td>No</td><td colspan="5">Yes</td></tr>
<tr><td colspan="2">Uniform Hazard Spectra</td><td>Yes</td><td>No</td><td colspan="4">Yes</td><td>No</td><td>Yes</td><td>No</td><td>Yes</td></tr>
<tr><td rowspan="3">Dis-aggregation</td><td>M-R</td><td colspan="4" rowspan="2">Yes</td><td rowspan="3">Yes</td><td rowspan="2">Yes</td><td rowspan="3">No</td><td rowspan="2">Yes</td><td>Yes</td><td>Yes</td></tr>
<tr><td>M-R-epsilon</td><td rowspan="2">No</td><td>No</td></tr>
<tr><td>Geographic</td><td colspan="4">No</td><td>Yes</td><td>No</td><td>No</td></tr>
<tr><td colspan="2">Eq. Scenario /Shake Maps</td><td>Yes</td><td>Yes</td><td colspan="5">No</td><td>Yes</td><td colspan="2">No</td></tr>
</table>

Fig. 6.1 Example of "skill matrix" for a set of software packages (From Danciu et al. 2010)

Recent studies have investigated and compared a number of available software in terms of their functionalities and features (e.g., Danciu et al. 2010) and in terms of hazard calculation results (e.g., Thomas et al. 2010; Bommer et al. 2013). A useful product of the study by Danciu et al. (2010) is what we may call a skill matrix of the PSHA programs (Fig. 6.1). This type of matrix may be explored to assess whether a particular code has the capabilities to address a specific issue of interest for the project (e.g., use of leaky or strict source boundaries in which a fault rupture initiating within a source is allowed to extend outside the source perimeter). In large projects, such as SIGMA, the PSHA software should be flexible enough to incorporate the results provided by the model developer on R&D actions initiated by the project or by the scientific community.

Important results are also provided by studies aiming at comparing the performance and accuracy of hazard results calculated using different codes for standard cases (see, e.g., Thomas et al. 2010). Such tests are extremely useful because they provide a well-defined benchmark for testing new codes and to evaluate the performances of existing codes for different situations. In the study by Thomas et al. (2010) the hazard curves calculated for a number of test cases using different software packages are compared among each other and, where available, with analytical solutions. In particular, the comparison with analytical solutions is extremely important, especially at very low probabilities of exceedance where numerical problems may arise in the codes.

The consistency of hazard results among different programs was only partially considered in SIGMA by performing a comparative study between Crisis and OpenQuake (D4-140). This study has the clear merit of pointing out that different hazard codes may use different modelling assumptions that result in different hazard estimates. It is thus important to verify that such assumptions are consistent with the PSHA model that is intended to be implemented. For example, the simulation of virtual ruptures within area source zones is clearly a topic in which different, and maybe equally reliable, models can be implemented (e.g., elliptical vs. rectangular ruptures). Furthermore, the study also pointed out the importance of the validation and verification protocol especially in terms of implementation of GMPEs. The approach followed in D4-140 is a good example of the strategy that should be followed when comparing hazard programs. The first step focuses on a comparison of the implementation between SIGMA and OpenQuake of the selected GMPEs for the French area. The second step concerns the comparison of hazard results obtained with CRISIS and OpenQuake for simple test cases. Finally, the CRISIS and OpenQuake hazard results are compared for a simplified version of the SIGMA PSHA logic tree.

6.3.2 Site Response Analysis Codes

Site response analyses are usually carried out under the assumption of one-dimensional wave propagation. Equivalent linear ground response modelling is by far the most commonly utilized procedure in practice (Kramer and Paulsen 2004) as it requires the specification of well-understood and physically meaningful input parameters (shear-wave velocity, unit weight, modulus reduction, and damping). Nonlinear ground response analyses provide a more accurate characterization of the true nonlinear soil behaviour, but they are seldom used in engineering practice because they require the specification of input parameters that lack an obvious association with fundamental soil properties and because the sensitivity of the site response results to these parameters is not well understood. The nonlinear codes differ essentially by the soil constitutive relationships built in the code and to a lesser extent by the numerical scheme: finite element (FEM), finite difference (FDM) or spectral method (SP).

The equivalent linear codes are robust and have been used extensively since the early seventies. Their main limitation lies in their inability to handle large amplitude motions because the embedded constitutive relationship (the equivalent linear model) is valid only for limited shear strains, typically of the order of twice the reference shear strain (see Sect. 5.1.4). Codes belonging to this category are SHAKE91 (Idriss and Sun 1992), EERA (Bardet et al. 2000) and CYBERQUAKE (Modaressi and Foerster 2000).

It is almost impossible and also not meaningful to list all possible nonlinear software. Several of them, belonging to all the categories listed above (FEM, FDM and SP), have been tested in the PRENOLIN benchmark (Deliverable SINAPS@-2015-V2-A2-T3-1) and in other benchmarks (Stewart et al. 2008). The most commonly used codes are DEEPSOIL (Hashash et al.), FLAC (ITASCA), SUMDES (Li et al. 1992), DYNAFLOW (Prevost 2010), OPENSEES (http://opensees.berkeley.edu/), NL-DYAS (Gerolymos and Gazetas 2005) and DMOD 2000 (Matasović and Ordóñez 2007). Only the last two softwares belong to the finite difference category. Once again, the merit of each code relies essentially on the capability of the soil constitutive model to reproduce the main features of the soil behaviour under cyclic loading but also, as evidenced in the PRENOLIN benchmark, on the qualification and experience of the users. An important lesson learned from SIGMA is the absolute necessity, before launching a large number of nonlinear calculations, of testing and validating the constitutive model against laboratory tests.

6.4 Sensitivity Analysis

During a project sensitivity analyses are used to test the impact of the different alternatives and to identify, for example, which parameters have a strong impact on the hazard at the site and may deserve further investigation. In practice, sensitivity analysis is often used during the course of a project to test alternative branches of the logic tree or to highlight the part of the model to which the SSC or the GMC teams should dedicate their efforts. Sensitivity analyses can be conducted to evaluate if a particular hypothesis should be included in the final logic tree or if its impact on the hazard at the site is negligible. In this latter case, the initial scientific logic tree can be simplified, thereby reducing the computational task for the project. Sensitivity analyses shall be done at a sufficiently early stage of the project to fully take advantage of the results. A sensitivity analysis generally aims at the following objectives:

- measuring the sensitivity of the seismic hazard to the different parameters/inputs (i.e., those that affect most the results); and
- evaluating the contribution of the uncertainties in the input parameters to the total hazard uncertainties.

Sensitivity analyses were performed for the French and Italian areas of interest in SIGMA (D4-18, D4-41, D4-29 and D4-94). A dedicated study is documented in D4-138 for the French area.

The uncertainties in the seismic hazard and the sensitivity of the hazard to the different parameters of the model can be measured using different metrics (D4-138):

- distance between hazard curve percentiles (e.g., 16–84%): this can be used, for instance, to compare the hazard from the whole logic tree among different sites or for different return periods. This metric can also be used to compare the hazard from a reduced logic tree, for example by comparing the distance between percentiles provided by the logic tree when using two area source models.
- ratio between percentiles (e.g. 16/50; 50/84) to appreciate the shape (symmetry) of the distribution.
- distance between mean hazard curves: this can be used, for instance, to compare the mean hazard curves obtained for different GMPEs.

The sensitivity analysis should be conducted for several return periods and several spectral periods (of interest for the project) in order to investigate the variation of the hazard distribution as a function of the annual frequency of exceedance at spectral periods consistent with the natural periods of the structure. This is done because typically the parameters that contribute to the hazard are not the same for short and long return or structural periods. Similarly, in low to moderate seismic regions, some parameters, e.g., the maximum earthquake magnitude, M_{max}, affect the seismic hazard more in the low-frequency range than in the high-frequency range; use of different spectral periods (e.g., PGA and SA at 0.2 s and 1 s) will serve that purpose.

In D4-138 the results of the sensitivity analysis are presented as Tornado diagrams, already introduced in Sect. 3.6.2. These diagrams constitute a compact graphical representation to summarize the results for a number of tests and for the two above mentioned metrics. Tornados are constructed first by fixing the return period of interest (e.g., 10,000 years) and finding in the mean hazard curves the acceleration corresponding to that return period, then the hazard variations can be appreciated in two ways:

- in terms of Annual Frequency of Exceedance (AFE) compared to the AFE of the mean acceleration; and
- in terms of spectral acceleration compared to the mean spectral acceleration. In this case the variability can be quantified as percentage variation around the mean acceleration value (Fig. 6.2).

Important information can also be obtained by examining the contribution of each source zone or fault included in the SSC logic-tree to the hazard at the site (this is also sometimes called disaggregation by source, although it is not a hazard disaggregation, strictly speaking). An example of hazard curves obtained from each source zone and their contribution to the total hazard at a site is presented in Fig. 6.3. This analysis allows highlighting which seismic sources carry the largest contribution to the hazard and thus may help to focus on better characterization of, e.g., seismicity of the zone, deformation mechanism and so forth.

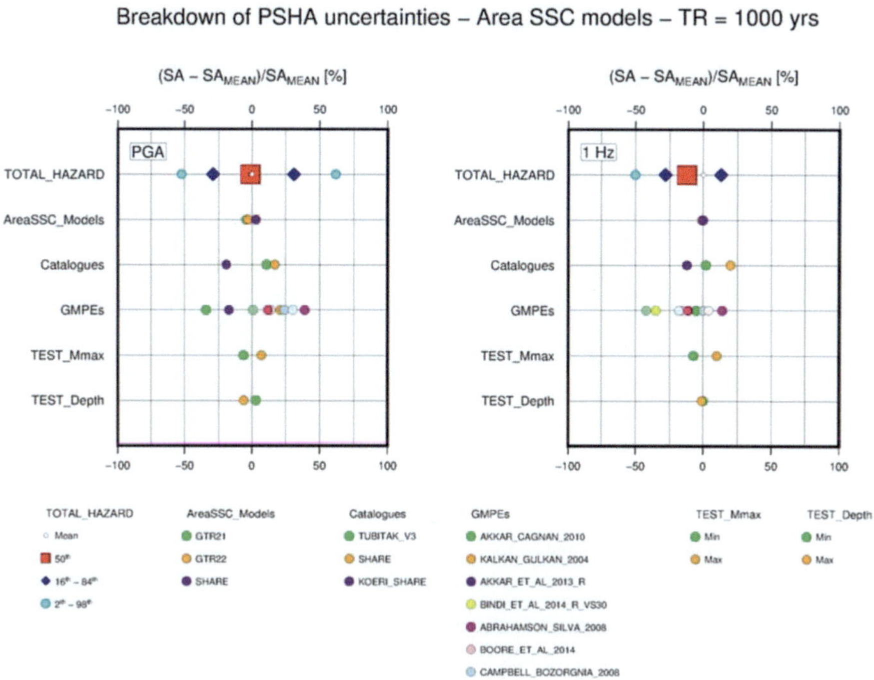

Fig. 6.2 Example of Tornado plot

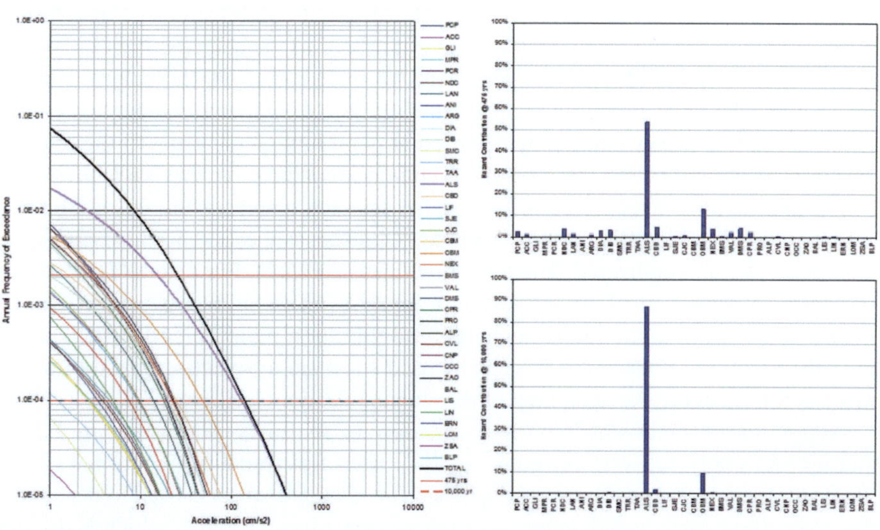

Fig. 6.3 Example of hazard contribution by source: hazard curves for each area source (*left*) and percentage contribution to the total hazard for two return periods of interest (*right*)

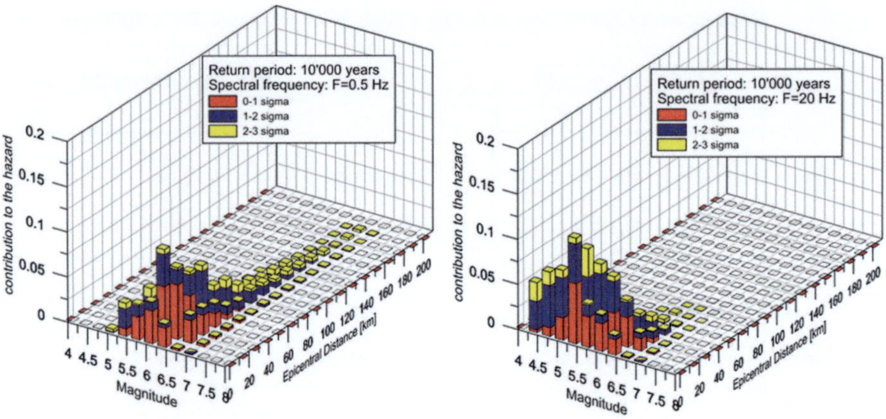

Fig. 6.4 Example of M-R-Epsilon (number of sigmas) hazard disaggregation for spectral acceleration at 0.5 Hz (*left*) and 20 Hz (*right*)

6.5 Hazard Disaggregation

The disaggregation of seismic hazard is an effective way to identify combinations of magnitude, distance, and values of other random variables (e.g., epsilon = the number of standard deviations of the ground motion distribution) defining seismic events that contribute to a selected seismic-hazard level (McGuire 1995; Bazzurro and Cornell 1999). The disaggregation process is carried out by evaluating the frequency of occurrence of seismic events with different combinations of magnitude, distance, and other random variables that contribute to the selected target seismic-hazard level (McGuire 1995) or that lead to exceedance of the selected target seismic-hazard level (Bazzurro and Cornell 1999). A higher frequency of occurrence of seismic events defined by a particular combination of values of the random variables describing possible seismic events is an indication that such identified events are the most likely to contribute to the selected target seismic-hazard level.

Therefore, the identified events can be used as scenario events for assessing both new and existing engineered structures and infrastructures. For example, if the target seismic-hazard level of interest represents a value of the structural design response spectrum at the fundamental natural vibration period of a designed structure, the identified scenario events can be used as an aid in selecting ground-motion records for checking the adequacy of the seismic design (see Sect. 6.7).

Magnitude-distance-epsilon disaggregation is routinely performed in PSHA studies and many software programs include such functionality. Figure 6.4 shows an example of *M*-*R*-epsilon disaggregation. The disaggregation is used to determine the controlling earthquakes in terms of magnitude, distance and ε, which is usually used as guidance to select or develop time histories for engineering purposes and at appropriate annual frequencies of exceedance. Epsilon disaggregation provides information on the contributing ground motions at the site in terms of the standard deviation of the GMPEs. This information may be useful to evaluate the coherence and consistency of the earthquake scenarios when deterministic seismic hazard assessment in conducted to control the PSHA results. It can also be used to depict the contribution of significant outliers to the total hazard, when the hazard appears to be controlled by the higher ε values.

6.6 Additional Engineering Output Parameters

Traditionally, PSHA is performed in terms of PGA and spectral acceleration (defined using the response of elastic single-degree-of-freedom (SDOF) systems). Alternative intensity measures (IMs) are proposed in the literature that attempt to be more closely related to the damage potential of an earthquake.

Some of them are amplitude-based (Peak Ground Velocity), duration-based [Arias Intensity, Cumulative Absolute Velocity (EPRI 1988), Standardized Cumulative Velocity (EPRI 1991)] or frequency response-based [Effective Peak acceleration (ATC 1978), Acceleration Spectral Intensity (Von Thun et al. 1988), Averaged pseudo-spectral intensity (Bojorquez and Iervolino 2011)].

In WP5, the SIGMA project investigated the respective merits and performance of existing and new proposed IMs for estimating the damage to structures and to non-structural components (NCS). A new simple and efficient intensity measure accounting for non-linear structural behaviour (De Biasio et al. 2014) and for probabilistic assessment of the acceleration demand of non-structural components (De Biasio et al. 2015) was proposed.

In a first step, a new IM, named Relative Average Spectral Acceleration (ASAR), is defined as a simple function of the fundamental frequency of the structure, of the reduced fundamental frequency of the damaged structure and of the spectral pseudo-acceleration between the two frequencies. Extensive tests, comprehensive comparative studies, statistical analyses based on recorded data from the RESORCE database (SIGMA WP2) and finite element simulations demonstrate that this new IM, with a reduction value R of 40% (ASA40), performs well compared to most of existing IMs and constitutes a promising engineering parameter to conduct performance-based seismic design. Based on a similar approach, a second parameter, namely Equipment-Relative Average Spectral Acceleration (E-ASA67) is proposed by De Biasio et al. (2015) to efficiently predict the acceleration demand to non-structural components.

With the objective to further implement the ASAR parameter in probabilistic risk assessment, a GMPE appropriate for the prediction of ASA40 was developed using the RESORCE strong ground motion database. Koufoudi et al. (2015) proposed a ASA40 prediction equation, with a domain of validity, while limited in the low magnitude range (M_w from 5.5 to 7.6), covers an appropriate distance range (<200 km) and spectral periods (0.01–4 s), that make this new GMPE a worthy candidate for future probabilistic seismic hazard and seismic risk analysis. However, these aspects were not implemented during the project.

6.7 Selection of Time Histories

6.7.1 Selection Based on UHS

Proper selection of input ground motions for linear and non-linear seismic analyses of structures and soil systems has become one of the leading topics of research in earthquake engineering. According to recent trends, as also recognized by the ASCE recommendations both for buildings and other structures (ASCE/SEI 7-10) and for nuclear power plants (ASCE/SEI 43-05), the set of input accelerograms should have some basic properties that may be summarized as follows:

- they should come from records of real earthquake events approaching, in terms of magnitude, distance and site conditions, the conditions that mostly affect seismic hazard for the specific return period;
- the average value of response spectra of the input accelerograms should closely approach the target UHRS (within a range that ASCE 43-05 recommends to be from −10% to +30%);
- a moderate scaling of accelerograms is generally accepted to improve the spectral matching with the UHRS; and
- such spectral matching in terms of average spectrum is recommended over a sufficiently wide period range to constrain not only the spectral ordinate at the fundamental period of the structure, but the spectral shape as well (e.g., Baker and Cornell 2006; Haselton et al. 2011).

For structural systems (see e.g., ASCE/SEI 7-10) the latter requirement is typically introduced to account for the sensitivity of nonlinear structural response to spectral ordinates at periods larger than the fundamental one (lengthening of the periods due to damage development), and for the higher modes contribution at shorter periods. For systems possibly involving nonlinear soil structure interaction effects, the requirement may even be more stringent, because soil response is governed by a wide range of frequencies and not by a narrow band around the fundamental period of the structure. In the latter case the ASCE/SEI 43-05 recommends considering the whole frequency range 0.2–25 Hz.

There is no standard method to select acceleration time histories to fit a prescribed target spectrum; the ideal case being the selection of real accelerograms recorded in seismic conditions close to the target ones, e.g., based on the magnitude, distance and epsilon coming from the disaggregation of the PSHA at rock. Different approaches have been proposed for such an optimum selection, although in practically all cases the accelerograms initially selected are more or less significantly modified by scaling procedures to improve the fit with the target spectrum.

In D3-96 for the site response analyses performed for sites in the Po plain, the approach for time-histories selection relies on the following ingredients (Smerzini et al. 2014):

- a high-quality strong motion database, containing digital records of earthquakes relevant for seismic hazard studies in Italy, obtained in stations with good seismic site characterization, in most cases provided by the V_{S30}; and
- a software tool for automatic selection of strong-motion records compatible on average with a target spectrum, within a prescribed period range and a prescribed tolerance.

In a second complementary step of the input motion selection procedure, a "spectrally matched" set of accelerograms is created. Starting from the previous selection, a scaling procedure in the frequency domain is applied. Namely, a correction factor for the Fourier spectrum of a prescribed accelerogram is calculated based on the ratio of the response spectrum (RS) of the accelerogram, with respect to the target one. Such a correction factor is applied to scale iteratively the amplitude of the Fourier spectrum of the accelerogram, while keeping the same phase as the original one, until its RS fits the target one. The procedure is similar to that proposed by Shahbazian and Pezeshk (2010).

Contrary to other spectral matching approaches, since the seed real accelerograms have a broadband average spectral compatibility and are also reliable in the long period range, the frequency scaling does not usually imply a significant modification of the original record, neither in terms of acceleration, nor of velocity or displacement.

In D4-153 different sets of time-histories were selected for linear, equivalent linear and nonlinear site response analyses at two selected sites. Two methods are applied for the selection and modifications. One set is selected following the approach employed in D3-96 above. The other procedure is much simpler and does not involve any spectral matching. The motions are searched within the RESORCE database (Akkar et al. 2014) according to magnitude-distance hazard disaggregation and site conditions. Then a linear scaling is performed such that five motions are scaled to match, in an average sense, the low-frequency (LF) part (above 0.2–0.3 s) of the target UHS and five are scaled to match the high-frequency (HF) part (below 0.2–0.3 s). The scaling factors are between 0.2 and 2.

As indicated in Sect. 5.4.2.1, the spectral matching technique has the advantage of reducing the uncertainty when broadband spectral matching of the records to the target spectrum is achieved. Furthermore, results obtained in SIGMA seem to indicate that when using SAFs to calculate ground surface response spectrum the choice of the set of time histories is not of paramount importance (Sect. 5.3) at least for the HyS approach.

6.7.2 Selection Based on Conditional Spectra

In recent years, ground-motion selection based on Conditional Spectra (Baker 2011; Jayaram et al. 2011; Carlton and Abrahamson 2014), has gained a growing relevance. In the framework of Seismic Probabilistic Safety Analysis (S-PSA) for nuclear power plants, the selection of relevant ground motions is the key step for defining the seismic load input for structural analysis. A classical method for structural analysis in S-PSA is the use of UHS, through generation of spectrum-compatible synthetic time histories. The conditional spectrum (CS) approach aims to avoid the conservatism of the UHS method. The use of such an approach has been investigated in D5-141.

From the point of view of the seismic hazard computation it is important to know which approach will be used to select the time histories in order to produce all the necessary results for its application. The main outputs typically provided are the UHS at the selected return periods and the magnitude-distance hazard disaggregation for the selected return periods and intensity measures (e.g., PGA and SA at 1 s). Hazard disaggregation is necessary to select time histories that are consistent with the source and propagation properties for the earthquakes that control the hazard at the site. If the CS approach is employed as in D5-141, the HCT needs to provide also hazard curves for each of the GMPEs used in the logic tree. This information is used, together with the GMPEs weights in the logic tree, to calculate disaggregation weights for each GMPE that correspond to the fractional contribution of that model to the total hazard for a given spectral period and hazard level.

Given the considerable amount of work involved in the CS method, compared to the other selection methods, and in view of the small sensitivity of the site responses (SAFs) to the choice of time histories, at least in the HyS methodology implemented in SIGMA, use of the CS method was not further pursued in the framework of SIGMA.

References

Akkar S,·Sandıkkaya MA, Senyurt M, Azari Sisi A,·Ay BÖ, Traversa P,·Douglas J, Cotton F, Luzi L,·Hernandez B, Godey S (2014) Reference database for seismic ground-motion in Europe (RESORCE). Bull Earthq Eng 12(1):311–339

Ameri G (2015) Progress Report - Preliminary Hazard Input Document (HID) for the SIGMA PSHA for the France's southeastern ¼, Doc. GTR/ARE/0915-1364, September 29, 2015, GEOTER, Clapier, France

Ameri G, Baumont D, Gomes C, Le Dortz K, Le Goff B, Martin C, Secanell R (2015) On the choice of maximum earthquake magnitude for seismic hazard assessment in metropolitan France – insight from the Bayesian approach. 9ième Colloque Nat. AFPS, Marne-La-Vallée, December 2015

American Society of Civil Engineers ASCE SEI 43-05 (2005) Seismic design criteria for structures, systems and components in nuclear facilities, New York, USA

American Society of Civil Engineers ASCE SEI 7/10 (2010) Minimum design loads for buildings and other structures, New York, USA.

ATC (1978) Tentative provisions for the development of seismic regulations for buildings. ATC-3-06 Report, Applied technology council, Redwood City, CA

Baker JW (2011) Conditional mean spectrum: tool for ground motion selection. J Struct Eng 137(3):322–331

Baker JW, Cornell CA (2006) Spectral shape, epsilon and record selection. Earthq Eng Struct Dyn 35(9):1077–1095

Bardet JP, Ichii K, Lin CH (2000) EERA – a computer program for equivalent-linear earthquake site response analyses of layered soil deposits, University of Southern California, Department of Civil Engineering

Bazzurro P, Cornell CA (1999) Disaggregation of seismic hazard. Bull Seism Soc Am 89(2):501–520

Bojorquez E, Iervolino I (2011) Spectral shape proxies and nonlinear structural response. Soil Dyn and Earthq Eng 31:996–1008

Bommer JJ, Strasser FO, Pagani M, Monelli D (2013) Quality assurance for logic-tree implementation in probabilistic seismic-hazard analysis for nuclear applications: a practical example. Seismol Res Lett 84(6):938–945

Carlton B, Abrahamson N (2014) Issues and approaches for implementing conditional mean spectra in practice. Bull Seism Soc Am 104(1):503–512

Danciu L, Monelli D, Pagani M, Wiemer S (2010) GEM1 hazard: review of PSHA software, GEM Technical Report 2010–2. GEM Foundation, Pavia

De Biasio M, Grange S, Dufour F, Allain F, Petre-Lazar I (2014) A simple and efficient intensity measure to account for non-linear structural behavior. Earthquake Spectra 30(4):1403–1426

De Biasio M, Grange S, Dufour F, Allain F, Petre LI (2015) Intensity measures for probabilistic assessment of non-structural components acceleration demand. Earthq Eng Struct Dyn 44(13):2261–2280

EPRI (Electric Power Research Institute) (1988) A criterion for determining exceedance of the operating basis earthquake EPRI NP-5930. Palo Alto, CA

EPRI, (Electric Power Research Institute) (1991) Standardization of the Cumulative Absolute Velocity. EPRI TR-100082, Palo Alto, CA

Gerolymos N, Gazetas G (2005) Constitutive model for 1-D cyclic soil behaviour applied to seismic analysis of layered deposits. Soils Found 45:147–159

Haselton C, Baker JW, Liel AB, Deierlein GG (2011) Accounting for ground motion spectral shape characteristics in structural collapse assessment through an adjustment for epsilon. J Struct Eng 137(3):332–344

Hashash YMA, Musgrove MI, Harmon JA, Groholski D, Phillips CA, Park D. DEEPSOIL V6.0. Urbana, IL

Idriss IM, Sun JI (1992) SHAKE91: a computer program for conducting equivalent linear seismic response analyses of horizontally layered soil deposits. Program modified based on the original SHAKE program published in December 1972 by Schnabel, Lysmer and Seed, Center of Geotechnical Modeling, Department of Civil Engineering, University of California, Davis, CA

International Atomic Energy Agency (2010) Seismic hazards in site evaluation for nuclear installations. Specific safety guide SSG-9, Vienna, Austria

Jayaram N, Lin T, Baker JW (2011) A computationally efficient ground-motion selection algorithm for matching a target response spectrum mean and variance. Earthquake Spectra 27(3):797–815

Koufoudi E, Ktenidou OJ, Cotton F, Dufour F, Grange S (2015) Empirical Ground-Motion Models adapted to a new intensity measure ASA40 (2015). Bull Earthq Eng 13(12):3625–3643

Kramer SL, Paulsen SB (2004) Practical use of geotechnical site response models. In: Proceedings of international workshop on uncertainties in nonlinear soil properties and their impact on modeling dynamic soil response, PEER Center Headquarters, Richmond, CA

Li XS, Wang ZL, Shen CK (1992) SUMDES: a nonlinear procedure for response analysis of horizontally-layered sites subjected to multi-directional earthquake loading. Department of Civil Engineering, University of California, Davis

Matasović N, Ordóñez G (2007) D-MOD2000, Geomotins, LLC, Computer software

McGuire RK (1995) Probabilistic seismic hazard analysis and design earthquakes: closing the loop. Bull Seism Soc Am 85:1275–1284

Modaressi H, Foerster E (2000) CyberQuake: modelling soil responses to seismic tremors. BRGM (http://www.brgm.eu/scientific-output/scientific-software/scientific-software)

NRC (2007) Regulatory Guide 1.208, A performance-based approach to define the site-specific earthquake ground motion, Washington, DC

Prevost JH (2010) DYNAFLOW: a nonlinear transient finite element analysis program, Princeton University, NJ

Shahbazian A, Pezeshk S (2010) Improved velocity and displacement time histories in frequency-domain spectral-matching procedures. Bull Seism Soc Am 100(6):3213–3223

Smerzini C, Galasso C, Iervolino I, Paolucci R (2014) Ground motion record selection based on broadband spectral compatibility. Earthq Spectra 30(4):1427–1448

Stewart J, On-Lei Kwok A, Hashash YMA, Matasovic N, Pyke R, Wang Z, Yang Z (2008) Benchmarking of nonlinear geotechnical ground response analysis procedures, PEER report 2008/04, Richmond, CA

Thomas P, Wong I, Abrahamson N (2010) Verification of probabilistic seismic hazard analysis computer programs. Pacific Earthquake Engineering Research Center; PEER 2010/106, Berkeley, CA

Von Thun JL, Rochim LH, Scott GA, Wilson JA (1988) Earthquake ground motions for design and analysis of dams, earthquake engineering and soil dynamics II – recent advances in ground-motion evaluation (GSP 20). ASCE, New York, pp 463–481

Chapter 7
Interfaces Between Subprojects

7.1 SSC and GMC Interfaces

Interface issues between the seismic source characterization and the ground motion characterization are usually implicitly handled by the hazard computation. Nevertheless, the most relevant interface topics, for example the depth distribution and the distance metric, should be discussed among experts from both sides of the interface. In the following sections, the interface issues identified as the most important for the hazard are discussed briefly.

The key generic interface issues for the SSC and GMC are listed and summarized below:

- Consistency of earthquake catalogue with ground-motion models:
  The magnitude scale used by the GMPEs (at present mainly M_w) needs to be consistent with the magnitude definition used in the earthquake catalogue and the magnitude-frequency distributions in order to avoid additional conversions with associated uncertainties. The magnitudes in the earthquake catalogue are mainly based on models of the attenuation of macroseismic intensities. The compilation of an intensity and ground-motion dataset for earthquakes with both types of data can be useful for this. Furthermore, although this approach has not been implemented in SIGMA, empirical correlations between the European Macroseismic Scale (EMS) and spectral acceleration (*Sa*) can be used to evaluate the regional variations in the EMS-Sa relations and provide an idea of the underlying uncertainties, which are part of the earthquake catalogue.
- Magnitude conversions:
  Modern earthquake catalogues use magnitude conversions to convert M_L into M_w and M_S into M_w or others. Attention must be paid to the consistency of M_w (and depth) derived from individual macroseismic data points (for historical earthquakes) with M_w derived from records (for instrumental earthquakes). This can be done from a limited set of calibration events in order to reduce the uncertainties

A. Pecker et al., *An Overview of the SIGMA Research Project*, Geotechnical, Geological and Earthquake Engineering 42, DOI 10.1007/978-3-319-58154-5_7

related to magnitude conversions and to limit the bias possibly introduced in the calculation of the activity rates. These conversions should also be checked against the conversion (if any) used in developing the selected GMPEs. More recent GMPEs are usually based on M_w so a conversion within the project is not needed, but there may have been conversions made to develop the databases for the GMPEs.

- Rupture dimensions:
  The magnitude-area scaling relations are used in the source characterization and in the finite fault simulations (FFS). The GMPEs are based on a subset of earthquakes with a given area(M) distribution. The area(M) scaling used by the SSC and GMC should be consistent. Modern databases include estimates of the rupture dimensions and can be used to quantify the area-magnitude relation implicit in the GMPEs. The area(M) scaling for the NGA dataset is consistent with the classical area(M) models, but the area(M) model used in the finite fault simulation is inconsistent with larger areas in the FFS; e.g. some FFS approaches have "unconventional" area(M) models which should be discussed between the SSC and GMC teams.
- Distance conversions:
  Distance conversions may be necessary depending on the source-to-site distance metric used by the selected GMPEs. Ideally, a distance conversion is avoided or is directly addressed in the hazard code using the native distance measures of the ground-motion models. For the reliability of the hazard estimates it is very important to have a consistent definition of distance metrics (see also Bommer and Akkar 2012). Today, with some GMPEs making use of the depth to the top of the rupture, this has become even more relevant as the hazard computed from near-field sources is sensitive to differences in the distance measures.
- Distance extrapolation:
  Even if rarely the case in European regions, it may happen that the SSC includes sources out to 500 km, while ground-motion models were usually only defined up to 200–300 km. In consequence, the ground-motion models need to be defined or extrapolated out to 500 km. Most of the time, this does not have a significant effect on the hazard, but should be addressed for consistency.
- Style-of-faulting and dip definitions:
  The SSC defines the style-of-faulting categories by dip whereas the GMC usually uses rake. These different definitions of style-of-faulting and dip need to be consistent. The range of dips in the ground-motion dataset should be checked for consistent with the dips for the SSC style-of-faulting classes. Experts can assign different weights to the different fault styles for each seismic source. Usually, dip angles are associated with these faulting styles in these particular seismic sources and uncertainty in this parameter can be introduced in the form of alternative dip angles. For example, it should be made clear whether "composite" fault mechanism (e.g., transtension) can be handled by the hazard software and how this is done. In the presence of zones with "unknown" style-of-faulting it should be decided how this will be modelled in the GM logic tree.

- Reducing M_{min}:
  In the past, most hazard calculations were done using a M_{min} of 5. Depending on the project requirements, the hazard calculation may want to use a minimum magnitude of 4.5 or even lower when a CAV filtering approach is applied or to check the PSHA outputs against observations in regions with moderate activity. For the SSC, there is no issue with reducing M_{min}. For the GMC, the available ground-motion models may not extrapolate well to smaller magnitudes. In this case, the ground-motion models will need to be adjusted for smaller magnitudes using the local ground-motion data as a constraint. The latest GMPEs [like NGA-West2 and some GMPEs developed within SIGMA WP2: Ameri (D2-131) and Drouet and Cotton (2015)] have been developed with data down to magnitude 3 and thus, can often be directly used, but the characteristics of the local ground-motion data should always be compared to the lower magnitude data used by the GMPE developers in order to identify potential regional differences.
- M_{max}:
  In a limited number of European seismic regions, the M_{max} values from the SSC can be as high as magnitude 8. The ground-motion models need to be checked regarding appropriate extrapolation to magnitude 8.
- Depth Distribution of Earthquakes in the Study Region
  The earthquake focal depth distributions in the study region should be determined, as they can be very different depending on the tectonic setting. This, of course, requires acquisition of high quality data over a sufficiently long time period. Depth is an important parameter for those GMPEs using rupture distance, as it affects the distance computation. Thus, this parameter should be discussed among the experts in charge of the SSC and GMC in order to have all the necessary background information to adopt a consistent definition.

  For the modelling, different magnitude-dependent depth distributions might be adopted, implicitly reflecting the uncertainty of this parameter. Thus, the hypocentral depth distribution can be: trapezoidal, triangular (a special form of trapezoidal distribution), truncated/non-truncated normal and multiple uniform (histogram) distributions. Similarly, a magnitude-dependent depth distribution may be appropriate where the largest events cannot occur in the shallower part of the seismogenic layer.

7.2 GMC and SRC Interfaces

Special attention should be paid to the treatment and transfer of uncertainties between the ground motion characterization on rock and the site/soil response characterization (SRC). This interface becomes relevant once the hazard evaluation is split into the evaluation of GMPEs only up to the bedrock level and then adding the site-specific soil response on top to reach a sub-surface layer or the surface. In case the GMPEs are evaluated directly up to the surface this interface, of course, does not exist in the described manner. In SIGMA both approaches have been investigated and one of the conclusions was that there is a potential benefit in separating the GMC on rock and the site-specific soil amplification in order to reduce in each area the associated uncertainties. The key issues that have to be addressed in order to avoid double counting of uncertainties are:

- Partition of aleatory variability;
- Host-to-target conversions and consistency with site amplification models;
- Maximum ground motion; and
- *V*/*H* models.

It should be mentioned that the maximum ground motion truncation models are handled differently when separating the rock and soil ground motions. For both, the same empirical database might be evaluated and used for the development of the individual models. In general, there are no problems. It is hard to define an upper limit to rock ground motion, especially in very hard rock conditions. On the other hand, there is more empirical data available for soil and based on the geotechnical soil properties, physical limits exist.

In the following, the common key interface issues between GMC and SRC are briefly discussed.

- A clear definition of the reference bedrock level and properties is necessary, as this is the handover between the rock hazard and the soil response. Seismologists see the bedrock as the geological base rock (generally the bottom of the sedimentary column), while engineers are used to the practical engineering bedrock as defined in construction standards. Furthermore, there are different interpretations of rock (e.g. soft, stiff, hard and very hard), which can be clarified by introducing, e.g., V_{S30} as a criterion. This also implies checking the consistency of the rock velocity profiles at the target site with the rock profiles for the GMPE datasets (see Chap. 4).
- For soft rock conditions, the ratio of spectral acceleration to PGA will reach 1 at about 40–50 Hz as opposed to hard rock sites, where this ratio is reached at about 100 Hz.
- Reducing M_{min} to 4.5 or lower requires the ground-motion models to be applicable through adjustments for $M < 5$ and that the extrapolation of non-linear effects (if any) to 4.5 or lower is checked.
- Avoiding double counting of site amplification variability due to input rock ground motions and uncertainty in the soil: The aleatory variability of site amplification (between sites with the same V_{S30} and within a single site for different input motions) is included in the GMC on rock if the traditional standard deviations for GMPEs are used. It is also included in the epistemic uncertainty of the site profile for the SRC and in the randomness of amplification from different input time histories. The focus here is on the variability due to the use of input time histories for rock ground motions applied at the bedrock level. There are some alternative approaches to address this potential double counting, which have been investigated and applied for SIGMA (see Sect. 5.4). Nowadays, there is a trend for having the GMC remove the epistemic part by switching to a

single-station sigma approach and having the SRC not add the aleatory variability from different input time histories for the linear range of input ground motions and to apply just the median amplification for site response. Furthermore, as done also in SIGMA, there is also the alternative to use time histories tightly matched to the rock UHS. Vertical aleatory variability: The GMC includes additional aleatory variability for the vertical ground motion components. The SRC needs to evaluate the value of this uncertainty so that site response experts can evaluate the effect on the vertical aleatory models. As the vertical component was not part of SIGMA, the latter has not been an issue within the project.

Other interface issues to be considered are

- The site response models incorporate non-linear effects. The ground-motion models used by the GMC also include non-linear effects. These non-linear effects should be consistent with the application. In case the reference rock conditions are greater than 700 m/s, the non-linearity in the ground-motion model will be very small. The differences in the non-linear models between GMC and SRC will, therefore, not have a significant effect.
- There are 2D and 3D effects in the site response models. Some of the variability due to 2D and 3D effects is captured in the GMC aleatory terms. It is not clear how much double counting exists due to this issue. This should be explored depending on the topographic and near-surface geological configurations in the study region.
- Frequency content of rock ground motions: The site response usually uses times histories to define the input rock motion. The consistency of the frequency content (spectral shape) of the Fourier spectra of the input rock motions used by SRC with that from the GMC model needs to be checked.
- Host-to-target parameter (Kappa) for input rock motion: The frequency content characterized by the GMC for the input rock motion implies some Kappa values. The site response input motions should be developed accordingly (see Sect. 4.3).

7.3 Single-Station Sigma

In terms of the interface between GMC and SRC, the use of single-station sigma removes the effect of site-to-site differences in the average amplification from the rock sigma, but the variability of the soil amplification (due to different rock input time histories, for example) is still included in the single-station rock sigma values. The common empirical datasets used to estimate sigma are dominated by ground motions in the linear range, which originated mainly from weak motions. So, the soil site amplification variability for linear response is mainly included in the rock site single-station sigma. When separating the evaluation of rock ground motions from soil (surface) ground motions the effects, if there is potential nonlinearity in the soil, need to be addressed. The site amplification variability for cases with highly nonlinear soil site response is not represented in the rock site single-station sigma.

Thus, if high levels of shaking within the soil are expected for the given site, an additional aleatory variability on top of the epistemic considerations can be included in the SRC part to cover those effects, or it can be explicitly covered in the SRC through consideration of nonlinear models.

The use of single-station sigma requires that the epistemic uncertainty in the site-specific amplification be captured in the SRC logic tree model through alternative models for the V_S profile and the non-linear properties, as illustrated already in Chap. 5.

This concept applies to both the horizontal and the vertical ground motion components.

7.4 V/H Models for Rock and Soil

Even though in SIGMA the vertical component of hazard was not treated, some considerations and issues are briefly summarized here for the interested reader.

In the GMC, the rock *V*/*H* ratios are evaluated and weighted separately from the horizontal median models. In the rock site hazard calculation, the *V*/*H* ratios are applied to the combined fractiles of the hazard curves for the horizontal component.

In the SRC, two approaches can be used to provide *V*/*H* ratios and amplification factors, which scale the horizontal rock ground motion to vertical soil ground motion: (1) *V*/*H* ratios for soil and amplification factors for horizontal motion are developed by the SRC experts and are combined, and (2) *V*/*H* ratios for rock by the SRC experts and amplification factors for vertical motion by the SRC experts are combined. For the first approach, the different soil *V*/*H* ratios are combined on a per-model basis with the model-specific horizontal motion amplification. For the second approach, the *V*/*H* rock ratios of all models are combined with the SRC vertical motion amplification factors. Both approaches cover horizontal-to-vertical motion scaling and amplification. The results of both approaches are combined into a single soil input for the target site. This soil input is used to compute the vertical soil motion hazard on the basis of the horizontal rock motion hazard, which is always the "full" (all fractiles and complete SSC and GMC models) horizontal motion rock hazard.

The second approach mentioned above is similar to the computation of the vertical motion rock hazard in that the full horizontal motion rock hazard is combined with all GMC *V*/*H* models, i.e. there is no correlation between (but full combination of) the GMC-specific rock hazard subsets and the GMC expert-specific *V*/*H* rock ratios.

References

Bommer JJ, Akkar S (2012) Consistent source-to-site distance metrics in ground-motion prediction equations and seismic source models for PSHA. Earthquake Spectra 28:1–15

Drouet S, Cotton F (2015) Regional stochastic GMPEs in low-seismicity areas: scaling and aleatory variability analysis – application to the French Alps. Bull Seismol Soc Am 105(4):1883–1902

Chapter 8
Probabilistic Seismic Testing and Updating of Seismic Hazard Results

Considering the high uncertainties in PSHA and the importance of PSHA results for seismic design and retrofit, it is pertinent to focus on the issue of consistency checking of the PSHA results. In the last decade several approaches to test PSHA results have been published (e.g. Stirling and Petersen 2006; Stirling and Gerstenberger 2009, 2010; Stirling 2012; Mucciarelli et al. 2008; Humbert and Viallet 2008; Labbé 2010; Anderson et al. 2011; Mezcua et al. 2013; Selva and Sandri 2013; Gribovszki et al. 2013). In addition, several recent opinion papers are encouraging hazard analysts to carry out tests of PSHA results (e.g. Stein et al. 2011). During the SIGMA period an international workshop was held in Pavia (NEA 2015) on "Testing PSHA results and Benefit from Bayesian Techniques for Seismic Hazard Assessment" which concluded that *a state-of-the-art PSHA should include a testing phase against any available observation, including any kind of observation and any period of observation, including instrumental seismicity, historical seismicity and paleoseismicity data if available. It should include testing not only against its median hazard estimates but also against their entire distribution (percentiles).*

Testing applications have been conducted in different countries (United States, New Zealand, Italy and France). The techniques rely on various statistical assumptions but any testing technique must deal with the available observation time window. Considering the observation time window in seismology testing (approximately 80 years for instrumental networks in California and 45 years in Europe, and several centuries for historical data), and the return periods of interest in engineering seismology (in the order of several hundreds to thousands of years), the comparison of observations and predictions is challenging.

Recent literature shows that PSHA models can be evaluated using different types of observations, such as intensities, synthetic accelerations (converted from intensities or predicted from an earthquake catalogue and a ground-motion model), and recorded accelerations at instrumented sites.

In the framework of SIGMA, WP4 specific actions were undertaken to develop new methods for comparing the ground-motion distribution with observations, in

A. Pecker et al., *An Overview of the SIGMA Research Project*, Geotechnical, Geological and Earthquake Engineering 42, DOI 10.1007/978-3-319-58154-5_8

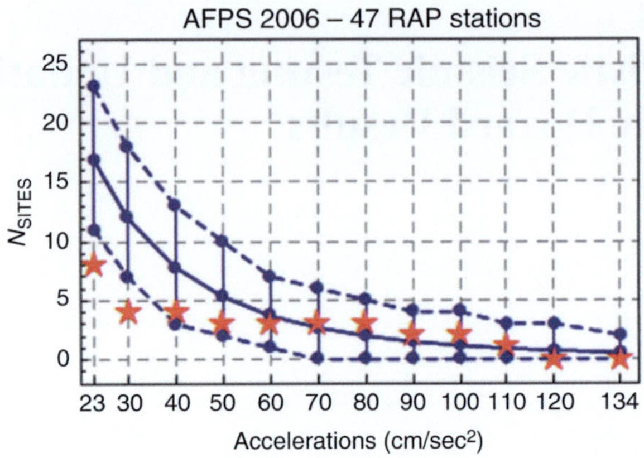

Fig. 8.1 Testing the probabilistic seismic hazard model against accelerometric data in France: predicted and observed number of exceedance. *Blue curves*: median and percentiles 2.5 and 97.5 of the predicted distributions, *red stars*: observed number of sites. Results for the AFPS2006 PSH model, considering a range of acceleration thresholds (47 rock sites, total time window 485 years) (From D4-75)

particular with the objective of implementing and testing methods to clearly understand their potential as well as their limits and accuracy.

Overall, the testing of PSHA models and results is an exercise that is worth pursuing.

8.1 PSHA Testing Using Acceleration and Macroseismic Intensity Data

The topic of testing probabilistic seismic hazard models against observations by using accelerometer data is discussed in SIGMA Deliverable D4-75. Three hazard models for France are tested against a set of French accelerometric records (RAP network, http://rap.resif.fr/). Two of the models (MEDD2002 and AFPS2006) provide hazard estimates for the whole metropolitan territory, while the third (SIGMA2012, D4-41) focuses on Southern France. The method is based on comparing predicted exceedance rates for a set of ground motion levels with observed ones at a set of sites (see Fig. 8.1). In order to reduce the influence of site conditions, which might considerably change the observed ground motion level, only recordings from rock sites are included in the analysis. Models are tested against different datasets, true-recorded accelerations and synthetic amplitudes based on an earthquake catalogue combined with a GMPE.

The PSHA testing using recorded accelerometer data presented in D4-75 is promising, although it suffers from a limited applicability to the French context due to the relatively moderate seismicity, which does not allow for testing significant ground-motion thresholds (i.e. low frequencies of exceedance). Another relevant issue is related to site response, which is typically not considered in the testing process by selecting only rock sites. However, the site classes of several French stations are not well known, which increases the uncertainties in the process.

The authors conducted an application of the same approach to Turkey in order to test PSHA for larger ground-motion levels (see Tasan et al. 2014). The SHARE hazard model (Woessner et al. 2015) is tested against Turkish strong-motion data. Two critical issues are identified: (1) the completeness of the recorded data at each station and the need to identify potential gaps in the recordings, and (2) the fact that most accelerometer stations are on soil sites while the hazard results are for rock sites. To solve this problem, PGA is converted into rock PGA using empirical site-amplification functions based on V_{S30} at the site. This process carries, however, large uncertainties that are not considered because V_{S30} is only a proxy of the real site amplification and only "average" site amplification can be estimated. The results show that the observed rates of exceedance at different sites are well within the bounds of the predicted distribution for accelerations between ~0.1 and 0.4 g. For higher levels no conclusion can be drawn. Despite the previous uncertainties, the method offers a rationale to assess a PSHA model at a 475-year return period.

In order to overcome the limitations related to the short lifetime of accelerometer stations, macroseismic intensity observations can be considered as an alternative, as they are available on a much longer period of time. Moreover, since macroseismic intensities are not directly used as input in standard PSHA models issues related to the independence between the model and the data used to test it may be avoided.

This approach is pursued in D4-118 and the follow-up study D4-154 that proposes a method to compare PSHA results with historical information on damage obtained from macroseismic intensity observations. The method consists in converting the intensities into PGA by means of fragility curves to obtain estimates of the PGA corresponding to the level of observed damage. Thus, the method requires a procedure to convert records of historical observations into equivalent PGA estimates. Such a conversion process carries uncertainties that are taken into account by means of a logic tree approach (see Fig. 8.2). The empirically-derived rates of exceedance are compared to the model derived estimates. The method was applied to south-eastern France.

The approach proposed in D4-118 suffers from very large uncertainties in the process of inverting the fragility functions. Such large variability in empirical PGA exceedance estimates prevents, in fact, the results of this methodology from being of practical use. In that respect, it should be mentioned that the methodology proposed by Labbé (2010) does not require any conversion from macroseismic intensities into accelerations nor does it require inversion of the fragility curves; however, this approach was not tested in SIGMA. A possible improvement, which was implemented in SIGMA, consists of including recorded accelerations in the procedure to better constrain the PGA distributions. In particular, this will improve the quality of estimation for lower values of PGA, whose rates of occurrence are likely to be underestimated by operating on the macroseismic database. Other important concerns are: (a) the treatment of the incompleteness of the catalogue, and the fact that

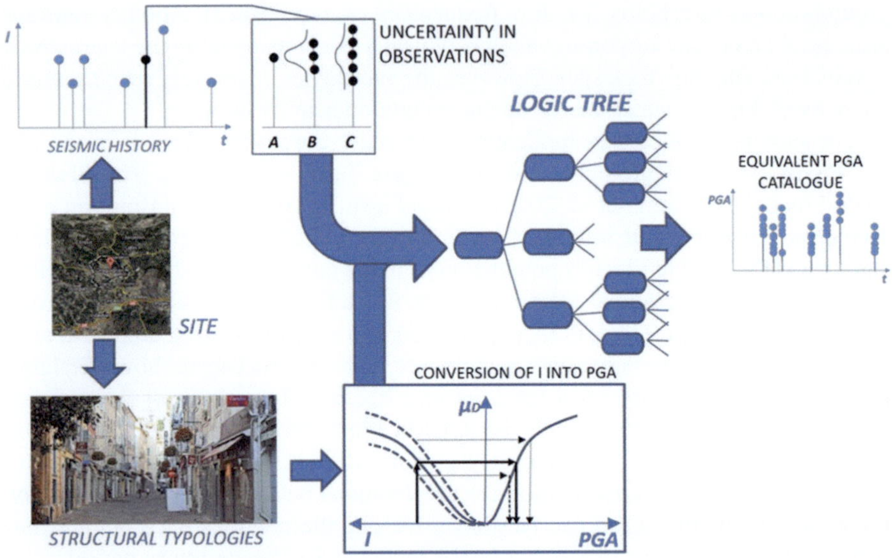

Fig. 8.2 Sketch of the proposed methodology in D4-118

large amount of damage is caused by induced events such as slope failures, and (b) the need to integrate site response in this method.

The limitations found in D4-118 were considered in D4-154. This continuation study was intended to respond to the criticisms and to explore several lines of development, the most important one being the use for comparison of the mean damage ratio (Lagomarsino and Giovinazzi 2006) rather than the ground motion itself; although this latter paper does not address directly PSHA testing it is essential as it ensures consistency between different fragility curves expressed as function of PGA or macroseismic intensity. The implementation of the revised method to the south-east quarter of France, and a complete re-visitation of the macroseismic database were also included in this work. The method is applied to single sites, aggregated sites, and at a regional scale. At each scale comparisons with historical data are carried out.

Results for single sites are not promising, as already discussed in D4-118. However, by using appropriate criteria in the selection of sites to satisfy the assumption of ergodicity (independence, and common distributions), the authors demonstrate that better comparisons are achieved for aggregated sites and for the regional scale.

The method proposed in D4-154 is judged to be promising. It will benefit from additional work to confirm its actual value as a tool to check the consistency of PSHA analyses. To gain confidence in the methodology, it was suggested in the project to consider a case study undertaken for a region with more abundant building damage data.

In conclusion, it is worth noting that the procedure proposed by Labbé (2010) does not require any conversion of macro seismic intensities into accelerations, not does it require inversion of the fragility curves.

8.2 Bayesian Update of PSHA

A different perspective in using observed data to test PSHA models is presented in D4-139. A Bayesian approach is applied to update the PSHA logic tree based on observed exceedances of a certain acceleration threshold. This approach differs from the previous ones because its objective is not to compare observed and predicted number of exceedances but to use such comparisons in a Bayesian framework in order to modify the logic tree weights. In practice, the logic tree weights of the branches that are in better agreement with the observations are increased.

The study performed in D4-139 presents a first simplified application of this approach. It provides the basis for the method and represents an initial step in examining ways of using additional information to update PSHA results in the French environment. The strategic choice of concentrating on updating the weights of the logic tree is appealing and could be made applicable to real cases; but it still requires additional research and development to reach such a stage.

Further improvements should be considered to extend this approach to a real and complete case study. In its present form, the following limitations and improvements are identified:

- The method considers only one ground-motion threshold level, therefore one set of new LT weights, to update the entire hazard curve. To be rigorous, since the set of updated weights changes with the value of the threshold, the updated set of weights should be calculated for every discrete value of the ground-motion threshold in the entire range of interest.
- The application of the method requires defining criteria to make an appropriate selection of the prior set of alternative branches, and an appropriate set of prior weights to the branches.
- The updating method should be applied in the spectral acceleration domain and not only to PGA. This would allow an appreciation of the variability of weights in the spectral domain and to update the full response spectrum.

References

Anderson JG, Brune JN, Purvance M, Biasi G, Anooshehpoor A (2011) Benefits of the use of precariously balanced rocks and other fragile geological features for testing the predictions of probabilistic seismic hazard analysis. In: Faber MH, Kehler J, Nishijima K (eds) Applications of statistics and probability in civil engineering. Taylor & Francis, London, pp 744–752

Gribovszki K, Bokelmann G, Szeidovitz Gy, Paskaleva I, Brimich L, Kovács K, Mónus P (2013) Comprehensive investigation of intact, vulnerable stalagmites to estimate an upper limit on prehistoric horizontal ground acceleration. Recent advances in earthquake engineering and structural dynamics (VEESD conference), 28–30 August, Vienna, Austria

Humbert N, Viallet E (2008) A method for comparison of recent PSHA on the French territory with experimental feedback. In: Proceedings of 14th world conference on earthquake engineering, 12–17 October, Beijing, China

Labbé P (2010) PSHA outputs versus historical seismicity; Example of France. In: 14th European conference on earthquake engineering, August 30–September 3, Ohrid, FYROM

Lagomarsino S, Giovinazzi S (2006) Macroseismic and mechanical models for the vulnerability and damage assessment of current buildings. Bull Earthq Eng 4(4):415–443

Mezcua J, Rueda J, García Blanco RM (2013) Observed and calculated intensities as a test of a probabilistic seismic-hazard analysis of Spain. Seismol Res Lett 84(5):772–780

Mucciarelli M, Albarello D, D'Amico V (2008) Comparison of probabilistic seismic hazard estimates in Italy. Bull Seismol Soc Am 98:2652–2664

NEA (Nuclear Energy Agency) (2015) Workshop on testing probabilistic seismic hazard analysis results and the benefits of Bayesian techniques, Pavia, Italy, 4–6 February, NEA/CSNI/R(2015)15. www.oecd-nea.org

Selva J, Sandri L (2013) Probabilistic seismic hazard assessment: combining Cornell-like approaches and data at sites through Bayesian inference. Bull Seismol Soc Am 103(3):1709–1722

Stein S, Geller R, Liu M (2011) Bad assumptions or bad luck: why earthquake hazard maps need objective testing. Seismol Res Lett 82(5):623–626

Stirling MW (2012) Earthquake hazard maps and objective testing: the hazard Mapper's point of view. Seism Res Lett 83(2):231–232

Stirling M, Gerstenberger M (2009) Development of tests for long-term earthquake ground motion forecasts in New Zealand, GNS science consultancy report 2009/140. p 20, Lower Hutt, NZ

Stirling M, Gerstenberger M (2010) Ground motion–based testing of seismic hazard models in New Zealand. Bull Seismol Soc Am 100(4):1407–1414

Stirling M, Petersen M (2006) Comparison of the historical record of earthquake hazard with seismic-hazard models for New Zealand and the continental United States. Bull Seismol Soc Am 96:1978–1994

Tasan H, Beauval C, Helmstetter A, Sandikkaya A, Guéguen P (2014) Testing probabilistic seismic hazard estimates against accelerometric data in two countries: France and Turkey. Geophys J Int 198(3):1554–1571

Woessner J, Laurentiu D, Giardini D, Crowley H, Cotton F, Grünthal G, Valensise G, Arvidsson R, Basili R, Dermicioglu MB, Hiemer S, Meletti C, Musson RW, Rovida AN, Sesetyan K, Stucchi M (2015) The SHARE consortium, the 2013 European seismic hazard model: jey components and results. Bull Earthq Eng 13(12):3553–3596

Chapter 9
Summary and Way Forward

During 5 years SIGMA has not only achieved significant steps forward in the methods for evaluating seismic hazard at a site, but it also contributed to establishing a strong network of academic institutions, researchers and engineering companies that will survive after the project ends. The overall organisation of the project fostered very lively and fruitful discussions between all participants, including with the members of the Scientific and Steering Committees.

Despite the amount of effort devoted to SIGMA, significant endeavours remain to be made to increase the reliability of seismic hazard predictions and to better understand and constrain the associated uncertainties. In fact, not all topics related to SHA have been covered during the course of the project, either because they were discarded after its start or because new topics emerged during the 5 year programme. Without repeating all the achievements and lessons learned, SIGMA clearly stressed the importance of collecting data to feed and constrain the models: a uniform seismic catalogue is essential to better characterize the seismic source models, the RESORCE database is useful for developing GPMEs adapted to the seismotectonic context of the considered regions and site characterization through field instrumentation is mandatory to classify the site, to evaluate the site response and to implement the single-station sigma approach. All these improvements in empirical knowledge contribute to a better estimation of the hazard and characterization of the epistemic uncertainties.

Aside from these achievements, the following topics were identified as important and warranting more studies in the future.

9.1 Seismic Source Characterization

More data and studies are needed to better characterize the source models and more specifically:

A. Pecker et al., *An Overview of the SIGMA Research Project*, Geotechnical, Geological and Earthquake Engineering 42, DOI 10.1007/978-3-319-58154-5_9

- In France fault characteristics should be studied to assess their geometries (dip angle, rake, dimensions) and to estimate their activity rates, in a way perhaps similar to the Italian DISS fault database;
- In areas of low to moderate seismicity the recurrence Gutenberg-Richter parameters, *a* and *b*, carry large uncertainties that should be better constrained, especially for magnitudes derived from macroseismic observations;
- The maximum magnitude associated with each source zone, which has a large impact on the hazard for long return periods, should be studied from different perspectives: a physically-based approach relying on scaling relationships related to fault dimensions, a statistically-based approach like the EPRI Bayesian method and a model-based approach in connection with the G-R parameters used to reflect the constraints imposed on the deformation rates. Ideally, these different approaches must be reconciled.

9.2 Ground Motion Characterization

- While an essential product of the SIGMA project has been the development of RESORCE, this database should be maintained, updated and enriched with additional parameters such as: fault rupture parameters, natural frequency of recording site f_0, κ, presence of basin effects and so on;
- Efforts to develop GMPEs specifically adapted to the European context should be pursued taking advantage of the RESORCE database;
- More research is needed to identify the most appropriate rock/hard rock adjustments to GMPEs, which is a problem typically encountered on exposed rock sites in SCRs;

9.3 Site Response Characterization

- Investigations on the nonlinear behaviour of soils under cyclic loading are needed from both the experimental and numerical viewpoint to answer critical questions such as: is nonlinear behaviour supported by experimental evidence? Beyond which threshold amplitude is it necessary to take it into account? What are the prerequisites and validation steps for a nonlinear constitutive model?
- The range of applicability of 2D geometric models for site response analyses must be ascertained;
- The spatial variability in soil characteristics needs to be considered in 2D site response evaluations;
- With the increasing number of available earthquake records, the selection of time histories for site response analyses should be guided in more detail;
- The evaluation of the vertical component of motion at the ground surface is an issue that deserves further investigation.

9.4 Hazard Calculation

- Utmost efforts should be devoted to testing of PSHA outputs. Testing should be performed with instrumental data for short return periods (~30 years), with historical seismicity in terms of macroseismic intensities for return periods up to some hundreds of years, and with paleoseismicity data for very long return periods (thousands of years). Testing can be made at a site scale where the hazard should be consistent with the local observations, and at a regional scale where the hazard should be consistent with the average regional hazard;
- Updating of PSHA using Bayesian approaches should be investigated, possibly beyond the simple updating of weights in the logic trees; however, even in this simple form, it helps assigning objective weights to the logic tree branches in place of the experts' judgment;
- Based on progress made concerning knowledge of faults, a totally different approach for hazard evaluation based on fault numerical simulations needs to be investigated. Such finite fault simulations can be developed independently of GMPEs and are mainly physics-based. Today, GMPEs are still sometimes used in this context for reference comparison or scaling.

9.5 Risk Assessment

These aspects have barely been touched upon in SIGMA and warrant additional studies.

- Identification of ground motion intensity parameters like ASA40 that could be good predictors of structural damage and subsequent development of associated GMPEs that would allow direct probabilistic risk assessment of critical facilities;
- Development of hazard consistent fragility curves for different typologies of structures (e.g. buildings, industrial facilities), systems and components.

Obviously, the issues listed above are far from representing an exhaustive list of all topics of interest, but they constitute open questions the answers to which would greatly contribute to improving hazard estimates and reducing the epistemic uncertainties attached to them.

Annexes

Annex 1: List of Committees Members

Steering Committee

Pierre LABBE, Chairman	EDF
Catherine BERGE-THIERRY	CEA
Paolo CONTRI	ENEL
Jean Pierre RZEPKA	CEA
Jean François SIDANER	AREVA

Senior Hazard Analysis Review Panel-SHARP

Yoshimitsu FUKUSHIMA	International Atomic Energy Agency (Japan)
Robert KENNEDY	RPK Structural Mechanics Consulting: (USA)
Roger MUSSON	British Geological Survey (UK)
Alain PECKER	Chairman of Scientific Committee (France)
Rui PINHO	GEM Foundation (Italy)

Scientific Committee

Alain PECKER, Chairman	French Association for Earthquake Engineering (France)
Pierre Yves BARD	ISTerre Institut des Sciences de la Terre, Grenoble (France)
David BAUMONT	FUGRO (France)

A. Pecker et al., *An Overview of the SIGMA Research Project*, Geotechnical, Geological and Earthquake Engineering 42, DOI 10.1007/978-3-319-58154-5

Olivier BELLIER	CNRS/Université Paul Cézanne, Aix-En-Provence (France)
Thierry CAMELBEECK	Royal Observatory of Belgium (Belgium)
John DOUGLAS[1]	BRGM Bureau de Recherche Géologique et Minière (France)
Ezio FACCIOLI	Politecnico di Milano (Italy)
Michel GRANET	EOST Ecole et Observatoire des Sciences de la Terre, Strasbourg (France)
Aybars GURPINAR	Consultant (Turkey)
Peter MOCZO	University of Bratislava (Slovakia)
Marco MUCCIARELLI[†]	University of Potenza (Italy)
Francesca PACOR	INGV Instituto Nazionale di Geofisica e Vulcanologia, Milano (Italy)
Roberto PAOLUCCI	Politecnico di Milano (Italy)
Philippe RENAULT	SWISSNUCLEAR (Switzerland)
Jean SAVY	SRC Savy Risk Consulting (USA)
Frank SCHERBAUM	University of Postdam (Germany)
Gianluca VALENSISE	INGV Instituto Nazionale di Geofisica e Vulcanologia, Rome (Italy)
Gordon WOO	Consultant (UK)

SIGMA Work Package Leaders

Gloria SENFAUTE	Project manager EDF (France)
Mirko CORIGLIANO	Deputy project manager ENEL (Italy)

WP1: Seismic Source
Leader: Kevin MANCHUEL (EDF)
Co-leader: Mirko CORIGLIANO (ENEL)

WP2: Prediction models
Leader: Paola TRAVERSA (EDF)
Co-leader: Fabrice COTTON (ISTerre)

WP3: Site effects
Leader: Fabrice HOLLENDER (CEA)
Co-leader: Deborah SICILIA (EDF)

WP4: Hazard Model
Leader: Jean-Michel THIRY (AREVA)
Co-leader: Gloria SENFAUTE (EDF)

WP5: Exploitation of Seismic Ground Motion
Leader: F. ALLAIN (EDF)
Co-leader: Stephane GRANGE (3SR, University Joseph Fourier of Grenoble)

[1] Now at University of Strathclyde, Glasgow, UK

SIGMA Partners

Basilicata University Potenza: http://www.unibas.it
BGS (British Geological Survey): www.bgs.ac.uk
BRGM (Bureau de recherches géologiques et minières): www.brgm.fr
C E R E G E (Centre Européen de Recherche et d'Enseignement): www.cerege.fr
Comenius University in Bratislava: http://www.uniba.sk
EMSM (European Mediterranean Seismological Centre): www.emsc-csem.org
EOST (Institut de Physique du Globe de Strasbourg): www.eost.u-strasbg.fr
EUCENTRE (European Centre for Training and Research in Earthquake): http://www.eucentre.it/
GEOAZUR: https://geoazur.oca.eu
GEOTER (Géologie Tectonique Environnement et Risques): www.geoter.fr
GEM (Global Earthquake Model), Italy: http://www.globalquakemodel.org/
IAEA (International Atomic Energy Agency): www.iaea.org
INGV (Istituto Nazionale di Geofisica e Vulcanologia, Milano: www.mi.ingv.it
INGV (Istituto Nazionale di Geofisica e Vulcanologia, Roma): www.ingv.it
IUSS (Institute for Advanced Study of Pavia): http://www.iusspavia.it/
ISTerre (Institut des Sciences de la Terre): www.isterre.fr
Laboratoire 3S-R (laboratoire Sols, Solides, Structures – Risques): www.3s.hmg.inpg.fr
METU (Middle East Technical University): http://www.metu.edu.tr/
ORB (Observatoire Royal de Belgique): http://www.astro.oma.be
POLIMI (Politecnico di Milano): www.polimi.it
RPK (Structural Mechanics Consulting)
SRC (Savy Risk Consulting): www.savyriskconsulting.com
SWISSNUCLEAR, Switzerland: http://www.swissnuclear.ch
UJF (Université Joseph Fourier): www.ujf-grenoble.fr
Université de Versailles Saint-Quentin-en-Yvelines: http://www.uvsq.fr
Université Paul Cézanne Aix Marseille: www.univ-cezanne.fr
University of California, Berkeley: http://www.berkeley.edu
University of Potsdam: http://www.uni-potsdam.de
Stéphane Drouet: Earthquake Ground-Motion Consultant
Virginia Tech University, USA: www.vt.edu

Annex 2: List of Publications

Publications in International Journals with Peer Review

Title	Authors	Review/Journal
Reference database for seismic ground-motion in Europe – (RESORCE)	Akkar S., Sandıkkaya M., Senyurt M., Azari Sisi A., Ay B.O., Traversa P., Douglas J., Cotton F., Luzi L., Hernandez L, Godey S.	Bull. Earthquake Eng., (2014), 12, 311–339
Taxonomy of kappa: a review of definitions and estimation approaches targeted to applications.	Ktenidou O., Cotton F., Abrahamson N., and Anderson J.	Seismol. Research Letters (2014), 85(1), 135–146.
*Comparisons among the five ground-motion models developed using RESORCE for the prediction of response spectral accelerations due to earthquakes in Europe and the Middle East	Douglas J., Akkar S, Ameri G., Bard PY, Bindi D., Bommer J., Singh Bora S., Cotton F.,·Derras B., Hermkes M.·Kuehn N.M.,·Luzi L.,· Massa M., Pacor F., Riggelsen C., Sandıkkaya A., Scherbaum F.,·Peter J., Stafford, Traversa P.	Bull. Earthquake Eng. (2014), 12, 341–358
Testing Probabilistic Seismic Hazard estimates against accelerometric data in two countries: France and Turkey.	Tasan H., Beauval C., Helmstetter C A., Sandikkaya A., Guéguen Ph.	*Geophys. J. Int.* (2014), 198, 1554–1571
*Towards fully data driven ground-motion prediction models for Europe	Derras B., Bard P.Y. Cotton F.	Bull. Earthquake Eng. (2014), 12, 495–516
Site Amplification: Dependency on V_{S30} and κ_0. Bull. Seism. Soc. Am.	Laurendeau A., Cotton F., Ktenidou O., Bonilla F., Hollender F.	Bull. Seism. Soc. Am. (2013), 103 (6), 3131–3148
Moderate Earthquake Teleseismic Depth Estimations: New Methods and Use of the Comprehensive Nuclear-Test-Ban Treaty Organization Network Data	Letort J., Vergoz J., Guilbert J., Cotton F., Sebe O., Cano Y.	Bull. Seism. Soc. Am. (2014), 104 (2), 593–607.
A simple and efficient intensity measure to account for non-linear structural behaviour	De Biasio M., Grange S., Dufour F., Allain F., Petre-Lazar I.	Earthq. Spectra, (2014), 30(4), 1403–1426
*Pan-European Ground-Motion Prediction Equations for the Average Horizontal Component of PGA, PGV, and 5%-Damped PSA at Spectral Periods up to 3.0 s using the RESORCE dataset	Bindi D., Massa M., Luzi L., Ameri G., Pacor F., Puglia R., Augliera P.	Bull. Earthquake Eng., (2014), 12, 391–530

Title	Authors	Review/Journal
The Ferrara thrust earthquakes of May-June 2012: preliminary site response analysis at the sites of the OGS temporary network	Priolo E., Romanelli, Barnaba C., Mucciarelli M., Laurenzano G., Dall'Olio L., Abu Zeid N., Caputo R., Santarato G., Vignola L., Lizza C., Di Bartolomeo P.	Annals of Geophysics, (2012), 55(4), 591–597
Recent evolution and challenges in the Seismic Hazard Analysis of the Po Plain region, Northern Italy	Faccioli E.	Bull Earthquake Eng (2013), 11(1), 5–33
*Simultaneous Quantification of Epistemic and Aleatory Uncertainty in GMPEs using Gaussian Process Regression	Hermkes M., Kuehn N. M., Riggelsen C.	Bull. Earthquake Eng., (2014), 12, 449–466.
*Fourier Spectral and Duration Models for the Generation of Response Spectra which are Adjustable to Different Source Propagation and Site Effects	Bora S. S., Scherbaum F., Kuehn N., Stafford P.	Bull. Earthquake Eng., (2014), 12, 467–493
A New Approach to Include Path Effects in Ground Motion Prediction Equations Using the M_w 9.0 Tohoku Earthquake Aftershocks.	Dawood H., Rodriguez-Marek A.	Bull. Seism. Soc. Am., (2013), 103(2B), 1360–1732.
Sensitivity of coda waves to lateral variations of absorption and scattering: radiative transfer theory and 2-D examples	Mayor J., Margerin L., Calvet M.	Geophysical Journal International, (2014), 197(2), 1–21.
Source of scatter in κ: analysis of Christchurch data and comparison with local ground motion prediction models	Van Houtte C., Ktenidou O.J., Larkin T., Holden C.	Bull. Seism. Soc. Am. (2014), 104 (4), 1899–1913.
Regional Stochastic GMPEs in Low-Seismicity Areas: Scaling and Aleatory Variability Analysis—Application to the French Alps	Drouet S., Cotton F.	Bull. Seism. Soc. Am. (2015), 105(4), 1883–1902
A Bayesian methodology applied to the estimation of earthquake recurrence parameters for seismic hazard assessment	Keller M., Pasanisi A., Marcilhac M., Yalamas T., Secanell R., Senfaute G.	Quality Reliability and Engineering Intern. (2014), 30(7), 921–933.
A new, improved and fully automatic method for teleseismic depth estimation of moderate earthquakes ($4.5 < M < 5.5$): application to the Guerrero subduction zone (Mexico).	Letort J., Guilbert J., Cotton F., Bondar I., Cano Y., Vergoz J.	Geophys. J. Int. (2015), 201(3), 1834–1848
Empirical Ground-Motion Models adapted to a new intensity measure (ASA40)	Koufoudi E., Ktenidou O.-J., Cotton F., Dufour F., Grange S.	Bull. of Earthq. Eng. (2015), 13 (12), 3625–3643

Title	Authors	Review/Journal
Uncertainty in PSHA related to the parametrization of historical intensity data	Mucciarelli M.	Nat. Hazards Earth Syst. Sci. (2014), 14, 2761–2765.
Intensity Measures for Probabilistic Assessment of Non-Structural Components Acceleration Demand	De Biasio M., Grange S., Dufour F., Petre Lazar I., Allain F.	Earthq. Eng. & Struct. Dyn. (2015), 44(13), 2261–2280.
3D numerical simulations of earthquake ground motion in sedimentary basins: testing accuracy through stringent models	Chaljub E.; Maufroy E.; Moczo P.; Kristek J.; Hollender F.; Bard P.-Y.; Priolo E.; Klin P.; De Martin F.; Zhang Z.; Chen X.	*Geophys. J. Int.* (2015) **201,** 90–111
Earthquake ground motion in the Mygdonian basin, Greece: the E2VP verification and validation of 3D numerical simulation up to 4 Hz	Maufroy E., Chaljub E., Hollender F., Kristek J., Moczo P., Klin P., Priolo E., Iwaki A., Iwata T., Etienne V., De Martin F., Theodoulidis N. P., Manakou M., Guyonnet-Benaize C., Pitilakis K., Bard P.-Y.	Bull. Seism. Soc. Am. (2015), 105 (3), 1398–1418
M_w estimation from crustal coda-waves recorded on analog seismograms	Denieul M., Sèbe O., Cara M., Cansi Y.	Bull. Seism. Soc. Am. (2015), 105(2A), 831–849.
*SiHex: Nouveau catalogue de sismicité instrumentale pour la France métropolitaine.	Cara M., Cansi Y., Schlupp A., Arroucau P., Béthoux N. et al	Bull. Soc. géol. France (2015), t. 186, no 1, 3–19
Evaluation of probabilistic site-specific seismic hazard methods and associated uncertainties, with applications in the Po plain, Northern Italy	Faccioli E., Paolucci R., Vanini M.	Bull. Seism. Soc. Am. (2015), 105 (5), 2787–2807
Tectonic and geomorphic analyses of the Belledonne border fault and its extension, Western Alps	Billant et al.	Tectonophysics (2015), 659, 31–52
On the Relationship between Fourier and 1 Response Spectra, 2 Implications for the Adjustment of Empirical Ground-Motion and 3 Prediction Equations (GMPEs)	Bora S.S., Scherbaum F., Kuehn N., Stafford P.	*Submitted for publication* Bull. Seism. Soc. Am.
Development of a Response 1 Spectral Ground-Motion Prediction Equation, 2 (GMPE) for Seismic Hazard Analysis from Empirical Fourier Spectral and 3 Duration Models	Bora S.S., Scherbaum F., Kuehn N., Stafford P., Edwards B.	Bull. Seism. Soc. Am. (2015), 105(4), 2192–2218.
The role of site effects in the comparison between code provisions and the near field strong motion of the Emilia 2012 earthquakes	Gallipoli M.R., Chiauzzi L., Stabile T.A., Mucciarelli M., Masi A., Lizza C., Vignola L.	Bull. Earthquake Eng. (2014), 12, 2211–2230

Title	Authors	Review/Journal
Understanding the physics of kappa and correlating it with site characterisation parameters: an application to Euroseistest.	Ktenidou O.-J., Drouet S., Cotton F., Abrahamson N.N.	Geophys. J. Int. (2015) **203,** 678–691
The Finite-difference Modelling of Earthquake Motions: Waves and Ruptures.	Moczo P., Kristek,J., Galis M.	ISBN 978-1-107-02881-4 Cambridge University Press
Squeezing kappa (κ) out of the Transportable Array – An example using band-limited data from Southern Arizona	Ktenidou O.J., Silva W., Darragh R., Abrahamson N.N., Kishida T.	*Submitted for publication* Bull. Seism. Soc. Am.
International benchmark on numerical simulations for 1D, non-linear site response (PRENOLIN): verification phase based on canonical cases.	Regnier J. et al	*Submitted for publication* Bull. Seism. Soc. Am.
*Semi-automated procedure for windowing time series and computing Fourier amplitude spectra (FAS) for the NGA-West2 database'.	Kishida T., Ktenidou O.-J., Darragh R., Silva W.	Pacific Earthquake Engineering Research Center, PEER report 2016/02.
*A methodology for the estimation of kappa (κ) for large datasets. Example application to rock sites in the NGA-East database and implications on design motions	Ktenidou O.-J., Abrahamson N.N., Silva W., Darragh R.	Pacific Earthquake Engineering Research Center, PEER report 2016/01.
InterPACIFIC project: comparison of Invasive and non-invasive methods for seismic site characterization. Part I: Intra-comparison of surface wave methods	Garofalo F., Foti S., Hollender F., Bard P.Y., Cornou C., Cox B.R., Ohrnberger M., Sicilia D., Asten M., Di Giulio G., Forbriger T., Guillier B., Hayashi K., Martin A., Matsushima S., Mercerat D., Poggi V., Yamanaka H..	J. of Soil Dyn. and Earthq. Eng. (2016), 82, 222–240.
InterPACIFIC project: comparison of Invasive and non-invasive methods for Seismic site characterization. Part II: inter-comparison between surface-wave and borehole methods.	Garofalo F., Foti S., Hollender F., Bard P.Y., Cornou C., Cox B.R., Dechamp A., Ohrnberger M., Perron V., Sicilia D., Teague D., Vergniault C..	J. of Soil Dyn. and Earthq. Eng. (2016), 82, 241–254.
Simulation of the Basin Effects in the Po Plain during the Emilia-Romagna Seismic Sequence (2012) using empirical Green's functions	Dujardin A., Causse M., Courboulex F., Traversa P.	Pure and Applied Geophysics (2016), 173, 1993–2010.

*Publications using the SIGMA data production or in connexion with scientific SIGMA actions but the research is not directly funded by SIGMA

Full Papers in Peer-Reviewed Conferences

Title	Authors	Symposium/Workshops
Relation between seismic ground motions and Structural damage	De Biasio M., Grange S., Dufour F., Petre-Lazar I., Allain F.	Symposium – SMiRT22, USA, August 2013
Nonstationary Stochastic Simulation of Strong Ground-Motion Time Histories: Application to the Japanese Database	Laurendeau A., Cotton F., Bonilla F.	Symposium – 15 WCEE, Lisboa, 2012
Validation of site-effect numerical predictions: New results from the Euroseistest Verification and Validation Project (E2VP)	Maufroy E., Bard P.-Y., Chaljub E., Hollender F., Kristek J., Moczo P., Theodoulidis N.	Symposium – 15 WCEE, Lisboa, 2012
Topographic effects in strong ground motion	Rai M., Rodriguez-Marek A., Yong A.	Symposium – 15 WCEE, Lisboa, 2012
A methodology for measuring the quality of probabilistic seismic hazard predictions in the SIGMA project framework	Bosco B., Thiry J.M., Rota M., Penna A., Magenes G., Senfaute G., Martin C.	Symposium – 15 WCEE, Lisboa, 2012
Correlation between seismic ground motions and floor acceleration demand for frame and structural-wall buildings	De Biasio M., Grange S., Dufour F., Petre-Lazar I., Allain F.	Symposium: Second European Conference on Earthquake engineering and Seismology. Istanbul, 2014
Engineering application of high frequency ground motions in Christchurch	Van Houtte C., Ktenidou O.-J., Larkin T., Holden C.	New Zealand Society for Earthquake engineering Technical Conference and AGM, Wellington, 26–28 Apr., 2013.
κ_0 estimates for rock stations in the NGA-West2 project	Ktenidou, O.-J., Kishida T., Darragh R., Silva W., Abrahamson N.N.	10th US National Conference on Earthquake engineering, Anchorage, 21–25 July, 2014

PhD Theses

PhD student	Title	Institution	Supervisors
Alain Dujardin	Seismic Ground Motion Prediction: site-specific hybrid approach	Geoazur (CNRS, Université de Nice Sophia Antipolis)	Courboulex F. (Geoazur) Causse M. (ISTerre)
Aurore Laurendeau	Sub-set of near-source earthquake record	ISTerre	Cotton F. (ISTerre) Hollender F. (CEA)
Hilal Oksuz	Testing probabilistic seismic hazard estimations against observations, application to France	ISTerre	Beauval C. (ISTerre) Guéguen P. (ISTerre)
Jean Letort	Characterization of moderate earthquake depths using teleseismic data.	ISTerre	Cotton F. (ISTerre)
Jérémy Billant	Study of a mega-structure: the Southwestern alpine front – Belledonne fault	CEREGE (Université Aix-Marseille)	Bellier O. (CEREGE) Hippolyte J.C. (CEREGE)
Marylin Denieul	Seismic Moment Magnitude and Crustal Wave Coda	EOST (Institut de Physique du Globe de Strasbourg)	Cara M. (EOST) Sèbe O. (CEA)
Thibault Fradet	Étude de la vulnérabilité des populations anciennes aux tremblements de terre en France 1650–1850	Université Observatoire de Versailles Saint Quentin en Yvelines	Quenet G. (UVSQ)
Jessie Mayor	Regional variation of seismic attenuation in Metropolitan France, Observation and Modeling	IRAP (CNRS, Université Toulouse III, Paul Sabatier)	Margerin L. (IRAP) Calvet M. (IRAP)
Marco De Biasio	Ground Motion Intensity Measures for Seismic Probabilistic Risk Analysis	3SR Grenoble	Grange S. (3SR Grenoble) Calvet M. (IRAP)
Haitham Dawood	Improving the understanding and methodologies of ground motion variability	Virginia Tech, USA	Rodriguez Marerk A. (Virginia Tech, USA)

List of Deliverables

D2-15	RESORCE: Seismic Motion Databank: 2011 Version Including Improvement of Meta-Parameters	S. Akkar (METU), F. Cotton (GFZ), S. Godey (CSEM), M. Senyurt, P. Traversa, (EDF)
D2-16	Pre Selection of Ground Motion Prediction Equations for France- S.	S. Drouet, (Geoter)
D2-17	Methodology and Testing of GMPEs Based on Ground Motion Data from France and Surrounding Countries	E. Delavaud, (ETH Zurich)
D4-18	Final Preliminary Probabilistic Hazard Map for France's Southeast ¼	D. Carbon, C. Gomes, Ch. Martin, (Geoter)
D4-19	Methods for Testing PSHA Against Observations (Accelerometer Data)	H. Oksuz, (ISTerre)
D1-27	Review of Active Faulting in the Po Plain	P. Burrato, F. Maesano, G. Valensise, (INGV)
D4-29	Preliminary PSHAs at Selected Sites, based on State of the Art Earthquake Source Models and Attenuation Relationships, Considering Both Outcropping Bedrock and Soil Conditions	E. Faccioli, (Polimi)
D2-33	Development of Regional Ground Motion Prediction Equations Covering a Wide Magnitude Range Based on Stochastic Simulations (Application to the Alps, Pyrenees and Rhine Graben Regions in France)	S. Drouet, (Geoter)
D2-34	The Use of Ground Motion Data and Macroseismic Intensities to Evaluate the Applicability of GMPEs to France	E. Delavaud, (Geoter)
D4-35	Methodological Approach for Measuring the Quality of Probabilistic Seismic Hazard Predictions	J.M. Thiry, B. Bosco, (Areva)
D3-36	Toward the Definition of Reference Ground-Motion	L. Laurendeau, (Isterre)
D3-38	Evaluation of The Relevance of the Numerical Simulation Approach for Site Effect Estimation	E. Maufroy, (Isterre)
D3-39	Inventory of Approaches to Account for Site Effects: Methodologies and Regulations	C. Lacave, M. Koller, (Resonance Ing. Sa)
D4-41	Initial Probabilistic Seismic Hazard Model for France's Southeast 1/4 –	D. Carbon, S. Drouet, C. Gomes, A. Leon, Ch. Martin, R. Secanel, (Geoter)
D1-49	Seismic Moment Magnitude and Crustal Wave Coda	M. Denieul, (Eost)
D1-50	Characterization of Earthquake Depth Using Teleseismic Data Developing New Methods for Moderate Events and Application to the Po Plain Earthquakes	J. Letort, (Isterre)

D1-51	Preliminary Earthquake Recurrence Models with Extensive Use of Site Data	M. Mucciarelli, (Università della Basilicata)
D5-52	Relation Between Seismic Ground Motion and Structural Damage and Function-Loss Part 1: Structural Wall Buildings – Preliminary Results	M. De Biasio, (3SR Grenoble)
D2-53	Ranking of Available GMPEs from Residual Analysis for Northern Italy and Definition of Reference GMPEs	F. Pacor, (INGV Milano)
D3-54	Evaluations of Seismic Ground Motion Variability at Soft Sites by 3d-1d Propagations Models, Including Christchurch and Selected Sites in The Po Plain	R. Paolucci, (Polit. Milano)
D1-66	Morphostructural Analysis of the Belledone Fault System (Western Alps)	J. Billant, (Cerege)
D1-67	Preliminary Earthquake Source Models of the Po Plain, Based on Full 3d Definition of Active Faults	G. Valensise, (INGV Roma)
D1-68	Methodological Approach in Analysing Historical Sources: A Way to Improve Vulnerability Studies	T. Fradet, (Université St Quentin)
D2-71	Stochastic GMPEs for France	S. Drouet, (Drouet Consulting)
D2-72	Calibration of GMPEs for Po Plain Region	F. Pacor, (INGV Milano)
D2-73	On the Magnitude Dependency of Ground Motion Decay with Distance	A. Dujardin, (Geosciences Azur)
D3-74	Preliminary Study on Empirical Approaches for Site Effect Characterization (Including in Situ Measurements)	M. Mucciarelli, (Università Della Basilicata)
D4-75	Development of Methods for Testing Probabilistic Seismic Hazard Estimates against Observations	H. Oksuz, (Isterre)
D1-89	Crustal Tomography 3D vs. 1D – Localisation in the Pyrenees	T. Theunissen, S. Cheverot, M. Sylvander, (OMP)
D2-91	Databank of Seismic Motion for Europe – RESORCE, 2013 Version: Improvement of Metaparameters/Inclusion of Weak Motion Data/ Internet Portal/Perspectives and Related European Initiatives	S. Akkar (METU), S. Godey (CSEM), J. Douglas (BRGM), L. Luzi (INGV), P. Traversa, (EDF)
D2-92	Metadata Uncertainties and Sensitivity of Empirical Ground Motion Prediction Equations Adapted to the French Context	G. Ameri, (Geoter)
D2-93	Ground Shaking Scenarios in the Po Plain with Special Emphasis on the Area Affected by the Earthquake Sequence of May 2012	R. Paolucci, (Polit. Milano)
D4-94	Site Specific Probabilistic Study for Po Plain Area	E. Faccioli, M. Vanini, R. Paolucci, (Polimi)
D3-96	Approaches to Account for Sites Effects in the PSHA of Selected Sites in The Po Area	R. Paolucci, E. Faccioli, C. Smerzini, M. Vanini, (Polimi)

D3-97	Characterization of Classes of Sites with a Large Potential to Cause Site-Effects Taking into Account the Geological Heterogeneities (Methodological Approach)	P. Moczo, (University of Bratislava)
D1-109	Revision of the Po Plain Earthquake Catalogue in Probabilistic Terms	M. Mucciarelli, (University of Potenza)
D1-110	Seismic Moment Magnitude and Crustal Coda Waves	M. Denieul, M. Cara, (Eost)
D2-111	Improving the Understanding and Methodologies of Ground Motion Variability	A. Rodriguez-Marek, (Virginia Tech)
D2-112	Kappa (κ) Physical Understanding and Estimation Approaches Targeted to Applications	O. Ktenidou, (Isterre)
D2-113	Spatial Variations of Seismic Attenuation in Metropolitan France from Observation and Modelling of the Seismic Coda	J. Mayor, (Irap)
D3-114	Prediction of Non-Linearity Using Numerical Simulation: Results of the Verification Phase of the PRENOLIN Program	J. Regnier, (Cerema)
D3-115	Characterizing Site Metadata of Accelerometric Network Stations: Results and Methodological Feedback from a RAP Station Survey	A. Dechamp, (Cea)
D3-116	Site Instrumentation Usefulness: Example of Application of Empirical Site Effect Estimation in Low Seismicity Context and Implementation Recommendations	V. Perron, (Cea)
D4-118	Estimation of Seismic Risks Using Structural Fragility Functions and Macroseismic Intensity Data	M. Rota, (Eucentre)
D5-119	Relation Between Seismic Ground Motion and Structural Damages & Function-Loss	M. Di Biasio, (3SR Grenoble)
D1-127	Seismotectonic and Geodynamic Synthesis of SE France	X. Sengelen, O. Bellier (Cerege), L Bollinger (Cea), K. Manchuel, (EDF)
D1-128	Study of the Relation "Isoseismal Areas" vs. "Depth-Magnitude", Based on Instrumental Data	A. Schlupp, M. Durrenberger, C. Sira, (EOST)
D1-129	Study of Instrumented Earthquakes that Occurred during the First Part of the 20th Century (1905–1962)	J. Benjumea, M. Cara, L. Rivera, (EOST)
D2-130	Development of Response Spectral Ground-Motion Prediction Equation (GMPE) from Empirical Fourier Spectral and Duration Models	S. Sigh Bora, (GFZ)
D2-131	Empirical Ground Motion Model Adapted to the French Context	G. Ameri, (Geoter)
D2-132	Ground Motion Uncertainty and Variability in Euroseistest, Greece	O. Ktenidou, (Isterre)
D2-133	Ground Motion Variability in the Po Plain Region	F. Pacor, (INGV Milano)

D3-134	Intercomparison of Methods for Site Parameter and Velocity Profile Characterization (INTERPACIFIC Project): Results and Learnings from the Blind-Test Stage and Suggestions for a "Good Practice Guide" for Non-Invasive Methods	S. Foti, F. Garofalo (Polito)
D3-137	3D Ground Motions Simulations for Site Effects Assessment: Learnings from Euroseistest Verification and Validation Project	E. Maufroi, (Isterre)
D4-138	Integration of Sigma Improvement for PSHA and Sensitivity Studies (Intermediate Results)	G. Ameri, (Geoter)
D4-139	Methodological Exercise for Testing a Statistical Approach to Update a Probabilistic Seismic Hazard Assessment	R. Secanell, C. Martin, (Geoter)
D4-140	Comparative Study on the PSHA Estimates Using Crisis and Openquake	M. Pagani (GEM Foundation Advisor), L. Rodriguez (UME School Pavia)
D5-141	Application of the Conditional Spectra (CS) Method	M. Marcilhac, K. Macocco, (Phimeca)
D1-147	Relation of Intensity Attenuation and Methodologies to Determine Seismological Parameters of Historic Earthquakes.	D. Baumont, (Fugro), J. Bonnet, (UNS Nice University)
D1-148	Catalogue of Isoseismal Areas for XXth Century French Historical Earthquakes (Io > VI)	J. Lambert, D. Monfort-Climent, O. Bouc, (BRGM)
V2A2T3-1 (D3-149)	PRENOLIN: Results from the Verification and Validation Phases & Present Recommendations for the Use of Non-Linear Simulation	J. Régnier, (Cerema), F.Bonilla (IFSTTAR)
V1A1T3-1 (D3-150)	Correction of Surface Records of their Site Effect before Developing GMPE: An Alternative Approach to Get Reference Incident Ground Motion (Application to Kik-Net Data)	F. Hollender, (CEA), L. Foundotos, A. Laurendeau, (CEA), PY. Bard, (Isterre)
D3-151	Identification of Key Site Features for Site Effect Evaluation: Extensive Numerical Sensitivity Studies of Sedimentary Basin Structures	P. Moczo, (Bratislava University)
D3-152	Good Practices Guidelines for Site Response Evaluation	
D4-153	SIGMA Integrated Exercise for Site-Specific PSHA	G. Ameri, (Geoter)
D4-154	Estimation of Seismic Risks Using Structural Fragility Functions and Macroseismic Intensity Data	A. Rosti, M. Rota, E. Fiorini, A. Penna, P. Bazzurro, G. Magenes, (Eucentre)
D1-168	Calibration of Macroseismic Attenuation Models for France	K. Manchuel, P. Traversa, (EDF), D. Baumont (Geoter)
D3-169	Good Practices Guidelines for Site Characterization	
D4-170	Probabilistic Seismic Hazard Assessment for South-Eastern France	C. Martin, G. Ameri, (Geoter) D. Baumont (Fugro)

Bibliography

Akkar S,·Sandikkaya MA, Bommer J (2014a) Empirical ground-motion models for point- and extended-source crustal earthquake scenarios in Europe and the Middle East Bull Earthq Eng 12(1):359–387

Akkar S,·Sandikkaya MA, Senyurt M, Azari Sisi A,·Ay BÖ, Traversa P,·Douglas J, Cotton F, Luzi L,·Hernandez B, Godey S (2014b) Reference database for seismic ground-motion in Europe (RESORCE) Bull Earthq Eng 12(1):311–339

Al Atik L, Kottke A, Abrahamson N, Hollenback J (2014) Kappa (κ) scaling of ground-motion prediction equations using an inverse random vibration theory approach. Bull Seismol Soc Am 104(1):336–346

American Society of Civil Engineers ASCE SEI 43-05 (2005) Seismic design criteria for structures, systems and components in nuclear facilities. American Society of Civil Engineers, New York

American Society of Civil Engineers ASCE SEI 7/10 (2010) Minimum design loads for buildings and other structures. American Society of Civil Engineers, New York

Anderson J, Hough S (1984) A model for the shape of the Fourier amplitude spectrum of acceleration at high frequencies. Bull Seismol Soc Am 74:1969–1993

Assatourians K, Atkinson G (2010) Verification of engineering seismology toolbox processed accelerograms: 2005 Riviere du Loup, Quebec earthquake. Available at www.seismotoolbox.ca

ATC (1978) Tentative provisions for the development of seismic regulations for buildings, ATC-3-06 Report. Applied Technology Council, Redwood City

Atkinson G, Boore D (2006) Earthquake ground-motion prediction equations for eastern North America. Bull Seismol Soc Am 96(6):2181–2205

Baker JW (2011) Conditional mean spectrum: tool for ground motion selection. J Struct Eng 137(3):322–331

Baker JW, Cornell CA (2006) Spectral shape, epsilon and record selection. Earthq Eng Struct Dyn 35(9):1077–1095

Bardet JP, Ichii K, Lin CH (2000) EERA—A computer program for equivalent-linear earthquake site response analyses of layered soil deposits. University of Southern California, Department of Civil Engineering, Los Angeles

Basili R, Valensise G, Vannoli P, Burrato P, Fracassi U, Mariano S, Tiberti MM, Boschi E (2008) The database of individual seismogenic sources (DISS), version 3: summarizing 20 years of research on Italy's earthquake geology. Tectonoph 453:20–43, database open for consultation at http://diss.rm.ingv.it/diss/

Bindi D, Pacor F, Luzi L, Puglia R, Massa M, Ameri G, Paolucci R (2011) Ground motion prediction equations derived from the Italian strong motion database. Bull Earthq Eng 9(6):1899–1920

A. Pecker et al., *An Overview of the SIGMA Research Project*, Geotechnical, Geological and Earthquake Engineering 42, DOI 10.1007/978-3-319-58154-5

Bindi D, Massa M, Luzi L, Ameri G, Pacor F, Puglia R, Augliera P (2014) Pan-European ground-motion prediction equations for the average horizontal component of PGA, PGV, and 5%-damped PSA at spectral periods up to 3.0 s using the RESORCE dataset. Bull Earthq Eng 12(1):391–430

Biro Y, Renault P (2012) Importance and impact of host-to-target conversions for ground motion prediction equations in PSHA. In: Proceedings of 15 world conference on earthquake engineering, Lisboa, Portugal

Bojorquez E, Iervolino I (2011) Spectral shape proxies and nonlinear structural response. Soil Dyn Earthq Eng 31:996–1008

Bommer JJ, Akkar S (2012) Consistent source-to-site distance metrics in ground-motion prediction equations and seismic source models for PSHA. Earthq Spectra 28:1–15

Bommer JJ, Douglas J, Scherbaum F, Cotton F, Bungum H, Fäh D (2010) On the selection of ground-motion prediction equations for seismic hazard analysis. Seismol Res Lett 81(5):783–793

Bommer JJ, Akkar S, Kale Ö (2011) A model for vertical-to-horizontal response spectral ratios for Europe and the Middle East. Bull Seismol Soc Am 101(4):1783–1806

Bommer JJ, Strasser FO, Pagani M, Monelli D (2013) Quality assurance for logic-tree implementation in probabilistic seismic-hazard analysis for nuclear applications: a practical example. Seismol Res Lett 84((6)):601, 938–945

Boore D (2003) Simulation of ground motion using the stochastic method. Pure Appl Geophysics 160:635–675

Boore D, Joyner W (1997) Site amplifications for generic rock sites. Bull Seismol Soc Am 87(2):327–341

Campbell K (2003) Prediction of strong ground motion using the hybrid empirical method and its use in the development of ground-motion (attenuation) relations in Eastern North America. Bull Seismol Soc Am 93(3):1012–1033

Carlton B, Abrahamson N (2014) Issues and approaches for implementing conditional mean spectra in practice. Bull Seismol Soc Am 104(1):503–512

Cauzzi C, Faccioli E (2008) Broadband (0.05 to 20 s) prediction of displacement response spectra based on worldwide digital records (2008). J Seismol 12(4):453–475

Cauzzi C, Dalguer L, Baumann C, Giardini D (2015a) Anatomy of near-field ground-shaking generated by dynamic rupture simulations. In: Conference present at SSA2015, Pasadena, CA, USA. Seismol Res Lett 862(2B):601

Cauzzi C, Faccioli E, Vanini M, Bianchini A (2015b) Updated predictive equations for broadband (0.01–10 s) horizontal response spectra and peak ground motions, based on a global dataset of digital acceleration records. Bull Earthq Eng 13(6):1587–1612

Comité Européen de Normalisation CEN (2004) Eurocode 8: design of structures for earthquake resistance—Part 1: General rules, seismic actions and rules for buildings. Comité Européen de Normalisation CEN, Brussels

Cotton F, Scherbaum F, Bommer JJ, Bungum H (2006) Criteria for selecting and adjusting ground-motion models for specific target applications: applications to Central Europe and rock sites. J Seismol 10(2):137–156

CPTI Working Group (2004) Catalogo parametrico dei terremoti italiani. Istituto Nazionale di Geofisica e Vulcanologia (INGV), Bologna, Italy. http://emidius.mi.ingv.it/CPTI04

Danciu L, Monelli D, Pagani M, Wiemer S (2010) GEM1 hazard: review of PSHA software, GEM Technical Report 2010-2. GEM Foundation, Pavia

De Biasio M, Grange S, Dufour F, Allain F, Petre-Lazar I (2014) A simple and efficient intensity measure to account for non-linear structural behavior. Earthq Spectra 30(4):1403–1426

De Biasio M, Grange S, Dufour F, Allain F, Petre LI (2015) Intensity measures for probabilistic assessment of non-structural components acceleration demand. Earthq Eng Struct Dyn 44(13):2261–2280

Delavaud E, Cotton F, Akkar S, Scherbaum F et al (2012) Toward a ground-motion logic tree for probabilistic seismic hazard assessment in Europe. J Seismol 16(3):451–473

DISS Working Group (2010) Database of Individual Seismogenic Sources (DISS), Version 3.1.01: a compilation of potential sources for earthquakes larger than M 5.5 in Italy and surrounding areas. http://diss.rm.ingv.it/diss/, © INGV 20092010 -Istituto Nazionale di Geofisica e Vulcanologia

Douglas J (2011) Ground-motion prediction equations – 1964–2010. www.gmpe.org.uk

EPRI (Electric Power Research Institute) (1988) A criterion for determining exceedance of the operating basis earthquake, EPRI NP-5930. Electric Power Research Institute, Palo Alto

EPRI, (Electric Power Research Institute) (1991) Standardization of the cumulative absolute velocity, EPRI TR-100082. Electric Power Research Institute, Palo Alto

Gerolymos N, Gazetas G (2005) Constitutive model for 1-D cyclic soil behaviour applied to seismic analysis of layered deposits. Soils Found 45:147–159

Haselton C, Baker JW, Liel AB, Deierlein GG (2011) Accounting for ground motion spectral shape characteristics in structural collapse assessment through an adjustment for epsilon. J Struct Eng 137(3):332–344

Hashash YMA, Musgrove MI, Harmon JA, Groholski D, Phillips CA, Park D (2015) DEEPSOIL V6.0. University of Illinois, Urbana

Idriss IM, Sun JI (1992) SHAKE91: A computer program for conducting equivalent linear seismic response analyses of horizontally layered soil deposits. Program modified based on the original SHAKE program published in December 1972 by Schnabel, Lysmer and Seed. Davis, Center of Geotechnical Modeling, Department of Civil Engineering, University of California

ITASCA. Consulting Group. FLAC, Explicit continuum modelling of nonlinear material behaviour. ITASCA, Minneapolis. http://www.itascacg.com/

Jayaram N, Lin T, Baker JW (2011) A computationally efficient ground-motion selection algorithm for matching a target response spectrum mean and variance. Earthq Spectra 27(3):797–815

Johnston AC, Coppersmith KJ, Kanter LR, Cornell CA (1994) The earthquakes of stable continental regions – assessment of large earthquake potential, Electric Power Research Institute (EPRI), TR-102261-V1, 2–1-98, Palo Alto, CA

Kottke A, Rathje E (2008) Technical manual for Strata, Report 2008/10. Pacific Earthquake Engineering Research (PEER) Center, Berkeley

Koufoudi E, Ktenidou OJ, Cotton F, Dufour F, Grange S (2015) Empirical ground-motion models adapted to a new intensity measure ASA40 (2015). Bull Earthq Eng 13(12):3625–3643

Kramer SL, Paulsen SB (2004) Practical use of geotechnical site response models. In: Proceedings of international workshop on uncertainties in nonlinear soil properties and their impact on modeling dynamic soil response. PEER Center Headquarters, Richmond

Ktenidou O, Gélis C, Bonilla L-F (2013) A study on the variability of kappa (κ) in a borehole: implications of the computation process. Bull Seismol Soc Am 103(2A):1048–1068

Li XS, Wang ZL, Shen CK (1992) SUMDES: a nonlinear procedure for response analysis of horizontally-layered sites subjected to multi-directional earthquake loading. Department of Civil Engineering, University of California, Davis

Matasović N, Ordóñez G (2007) D-MOD2000, Geomotins, LLC, Computer software

Modaressi H, Foerster E (2000) CyberQuake: modelling soil responses to seismic tremors. BRGM. http://www.brgm.eu/scientific-output/scientific-software/scientific-software

Nagashima F, Matsushima S, Kawase H, Sanchez-Sesma FJ, Hayakawa T, Satoh T, Oshima M (2014) Application of horizontal-to-vertical spectral ratios of earthquake ground motions to identify subsurface structures at and around the K-NET site in Tohoku, Japan. Bull Seismol Soc Am 104(5):2288–2302

Poggi V, Edwards B, Fäh D (2012) Characterizing the vertical-to-horizontal ratio of ground motion at soft-sediment sites. Bull Seismol Soc Am 102(6):2741–2756

Prevost JH (2010) DYNAFLOW: a nonlinear transient finite element analysis program. Princeton University, Princeton

Rodriguez-Marek A, Montalva GA, Cotton F, Bonilla F (2011) Analysis of single-station standard deviation using the KiK-net data. Bull Seismol Soc Am 101(3):1242–1258

Secanell R, Martin C, Viallet E, Senfaute G (2015) A Bayesian methodology to update the probabilistic seismic hazard assessment. In: CSNI workshop on "testing PSHA results and benefit of Bayesian techniques for seismic hazard assessment", 4–6 February 2015, Eucentre Foundation, Pavia, Italy

Shahbazian A, Pezeshk S (2010) Improved velocity and displacement time histories in frequency-domain spectral-matching procedures. Bull Seismol Soc Am 100(6):3213–3223

Smerzini C, Galasso C, Iervolino I, Paolucci R (2014) Ground motion record selection based on broadband spectral compatibility. Earthq Spectra 30(4):1427–1448

Stewart J, On-Lei KA, Hashash YMA, Matasovic N, Pyke R, Wang Z, Yang Z (2008) Benchmarking of nonlinear geotechnical ground response analysis procedures, PEER Report 2008/04. Pacific Earthquake Engineering Research Center, Richmond

Thomas P, Wong I, Abrahamson N (2010) Verification of probabilistic seismic hazard analysis computer programs, Pacific Earthquake Engineering Research Center; PEER 2010/106. Pacific Earthquake Engineering Research Center, Berkeley

Van Houtte C, Drouet S, Cotton F (2011) Analysis of the origins of κ (kappa) to compute hard rock to rock adjustment factors for GMPEs. Bull Seismol Soc Am 101(6):2926–2941

Von Thun JL, Rochim LH, Scott GA, Wilson JA (1988) Earthquake ground motions for design and analysis of dams. In: Earthquake engineering and soil dynamices II – Recent advances in ground-motion evaluation (GSP 20). ASCE, New York, pp 463–481

Zhao J, Zhang J, Asano A, Ohno Y, Oouchi T, Takahashi T, Ogawa H, Irikura K, Thio HK, Somerville PG, Fukushima Y (2006) Attenuation relations of strong ground motion in Japan using site classification based on predominant period. Bull Seismol Soc Am 96(3):898–913

Index

A. Pecker et al., *An Overview of the SIGMA Research Project*, Geotechnical, Geological and Earthquake Engineering 42, DOI 10.1007/978-3-319-58154-5

MIX
Papier aus verantwortungsvollen Quellen
Paper from responsible sources
FSC® C105338

If you have any concerns about our products,
you can contact us on
ProductSafety@springernature.com

In case Publisher is established outside the EU,
the EU authorized representative is:
Springer Nature Customer Service Center GmbH
Europaplatz 3, 69115 Heidelberg, Germany

Printed by Libri Plureos GmbH
in Hamburg, Germany